Redox-active Polymers for Energy Storage and Medical Devices

Editor

Narendra Pal Singh Chauhan

Department of Chemistry
Faculty of Science
Bhupal Nobles' University, Udaipur
Rajasthan, India

CRC Press
Taylor & Francis Group
Boca Raton London New York

CRC Press is an imprint of the
Taylor & Francis Group, an **informa** business
A SCIENCE PUBLISHERS BOOK

Cover credit: Image created by Dr. Atefeh Derakhshani, Tehran University of Medical Sciences, Tehran, Iran.

First edition published 2026
by CRC Press
2385 NW Executive Center Drive, Suite 320, Boca Raton FL 33431

and by CRC Press
4 Park Square, Milton Park, Abingdon, Oxon, OX14 4RN

CRC Press is an imprint of Taylor & Francis Group, LLC

Library of Congress Cataloging-in-Publication Data (applied for)

ISBN: 978-1-032-87520-0 (hbk)
ISBN: 978-1-032-90242-5 (pbk)
ISBN: 978-1-003-54667-2 (ebk)

DOI: 10.1201/9781003546672

Typeset in Times New Roman
by Prime Publishing Services

Preface

Redox-active polymers are at the forefront of innovation, offering promising solutions in both energy storage and medical technologies. These materials, known for their ability to undergo reversible redox reactions, are paving the way for more efficient and sustainable batteries, as well as groundbreaking advancements in medical devices.

This book, Redox-active Polymers for Energy Storage and Medical Devices, provides a comprehensive yet focused exploration of these cutting-edge materials. It covers a range of topics, including the fundamental principles of redox polymers, their classification, and their diverse applications. The book delves into the development and use of conjugated and non-conjugated redox polymers, highlighting their role in enhancing the performance of energy storage systems. Additionally, it examines specialized polymers such as anthraquinone, catechol, phenothiazine, and ferrocene-based polymers, which are tailored for specific applications in organic batteries and medical devices. The book concludes with an exploration of composite polymer electrolytes, emphasizing their significance in improving the efficiency and reliability of next-generation batteries.

This book is intended for researchers, students, and professionals across fields such as polymer chemistry, materials science, energy technology, and biomedical engineering. Whether you are looking to gain a foundational understanding or seeking to advance your research, this book provides a well-rounded resource that bridges the gap between theory and practical application. I hope that the insights shared here will inspire further innovation and contribute to the continued growth of this exciting field.

Narendra Pal Singh Chauhan

Contents

Chapter 1

Radical-based Redox Polymers for Rechargeable Batteries

Navaneeth Kumar Ravikumar,[1] *Panneerselvam Perumal,*[1,*]
Manjinder Kour[2] and *Narendra Pal Singh Chauhan*[3]

1. Introduction

The dependence of our technology-driven, mobile society on energy storage has spurred advancements in the field of fuel cells. Fuel cells are being used in motor vehicles (Brandon et al., 2003), supercapacitors (Inagaki et al., 2010), and batteries (Linden David, 2002), among other applications. A sustainable secondary energy source should be affordable, abundant, reusable, and suitable for lighting applications, such as flashlights, smart cards, and innovative food packaging. Additionally, there is a need to design high-capacity, lightweight rechargeable batteries for use in laptops and electric cars. Smart grids and energy storage systems are also essential for replacing conventional power plants with renewable energy sources. While the basic principles of batteries are relatively simple, advancements in electric energy storage have not progressed as rapidly as in other areas of electronics (Akinyele and Rayudu, 2014; Hameer and van Niekerk, 2015). The average increase in rechargeable battery energy density is just 3% over the last 60 years (Zu and Li, 2011). The discovery of the lithium-ion battery marked a significant milestone in the ongoing pursuit of environmentally friendly energy storage devices. These batteries offer much higher specific energy and power densities compared to lead-acid batteries, making them suitable for a wide range of modern applications (Andersson and Råde, 2001). In

[1] Department of Chemistry, SRM Institute of Science and Technology, Kattankulathur, Chengalpattu, Tamil Nadu, India. 603203.
[2] Department of Chemistry and Biochemistry, Montana State University, Bozeman, Montana, USA.
[3] Department of Chemistry, Bhupal Nobles' University, Udaipur 313002, Rajasthan, India.
* Corresponding author: panneerp1@srmist.edu.in, panneerchem82@gmail.com
DOI: 10.1201/9781003546672-1

2007, Japan produced nearly 6 billion battery cells worth approximately $8 billion, grabbing almost half of the worldwide production of lithium-ion batteries (Fukao et al., 2007). Among the solutions provided to the problems, this electrochemical energy storage system offers advantages such as easy storage and transport, high conversion efficiency, and a great ability to minimize power fluctuations (Whittingham, 2004). In the last few decades, metal-ion batteries have emerged with different metals including Li^+, Na^+, K^+, Zn^{2+}, Mg^{2+}, Ca^{2+}, Al^{3+}, etc. (Lu et al., 2018). These have gained much attention because of their high energy density and are the most widely used electrochemical energy storage devices. However, the electrode materials comprising these batteries are usually made from conventional transition metals, such as Co, Ni, Mn, V, etc. Furthermore, the cycling process, where ions intercalate into and out of inorganic materials, always results in volume expansion, phase transitions, and even pulverization, leading to capacity decay due to the inflexibility of the bonds (Chen et al., 2008).

Reliable and eco-friendly rechargeable metal-ion batteries have recently gained importance, especially due to their higher energy and power densities and good stability. Conjugated polymers are a very attractive advantage over inorganic electrode materials, such as $LiCoO_2$ (Qiu et al., 2012), $Li_4Ti_5O_{12}$ (Chen et al., 2001), etc. This makes them very energetic in terms of energy storage; in terms of energy released, their conjugation is extended and the organic structure is varied, including addition of new functional groups or redox moieties (Song and Zhou, 2013). The employment of redox-active conjugated polymers enables quick and easy completion of the corresponding reagent, effectively preventing structural changes that are common for inorganic materials. These merits enable the use of conjugated polymers to achieve exceptional rate performance and remarkably long cycle life in rechargeable batteries (Häupler et al., 2015). Metal-ion technology is featured today among the most broadly used rechargeable battery systems. Lithium ions are of special interest in this, as they have exceptionally high energy and power density. Successful applications of organic materials as positive or negative electrodes for rechargeable Li-ion, Na-ion, and dual-ion batteries are presented (Novák et al., 1997). Thus, this point of view will address these battery systems primarily. This view focuses on important developments in nanostructured conjugated polymers. We begin with a discussion on some basics of conjugated polymers, followed by the more recent developments in the same field (Xie and Zhang, 2016). Further, we discuss the challenges in enhancing the efficiency of nanostructured conjugated polymer electrodes and suggest some techniques for enhancing their performance.

Redox polymers are compounds that can accept or lose an electron while undergoing oxidation through electron loss or reduction by electron gain. These types of polymers contain electroactive sites or groups through which they undergo redox reactions: either along the backbone, for example, like electrically conducting polymers and polypyrrole, or as pendant groups in those polymers possessing ferrocene groups (MacDiarmid, 2001; Pietschnig, 2016). The reaction process changes the oxidation state of polymeric material, further changing its properties. The chemical nature of the polymer will be different in terms of mechanical properties whether it has been reduced or oxidized. An organic redox-active material-based next-generation energy storage system is crucial and contributes to substantial superiority in properties, like flexible molecular structure, enhanced chemical and physical properties, and greater electrochemical storage performance compared with their metal-based inorganic counterparts. Organic materials would thus provide novel perspectives on existing energy storage technologies and create a diverse environment for the further development of new energy systems (Huo et al., 2014; Janoschka et al., 2012; Oyaizu and Nishide, 2009; Sui et al., 2013). This is even more relevant because redox-active polymers are now much more common than their organic small molecule analogs with equivalent reagent functions. Polymeric materials also inherently enjoy the advantages of improved stability, better process efficiencies, and easier device fabrication scenarios that look very promising for the eventual emergence of redox-active polymers as a potential candidate for future energy storage technology. The formation of radical-based redox polymers (RRP) is indicative of a new class of organic materials and is distinguished by stable organic radicals as their redox centers. These polymers have gained significant attention in recent years due to their unique advantages over conventional inorganic materials used in battery electrode manufacture. Inorganic systems depend on intercalation and alloying mechanisms for their functionality, while radical-based polymers store and release charge primarily through redox processes involving organic radicals (Muench et al., 2016; Shi and Yu, 2016). The mechanism shows impressive fast charging and discharging rates, along with excellent cycle stability and favorable power densities, indicating a highly efficient reaction process. One of the key advantages of radical-based redox polymers (RRP) is their design flexibility. This flexibility allows for modifications in their redox properties, solubility, and mechanical characteristics, enabling them to meet specific battery requirements.

As a result, this book chapter will focus on recent advances in the area of radical-based redox polymers in terms of study, characterization, and use. This chapter endeavours to give a comprehensive explanation of radical-based redox polymers as an emerging alternative for rechargeable batteries. The main study areas entail the basics of radical-based redox polymers, electrochemical properties of RRPs in rechargeable batteries, their designs and engineering into better batteries, challenges, limitations, possibilities, and applications. The chapter provides an overview of the role of radical-based redox polymers as sustainable, high-performance materials for rechargeable batteries, ideal for the researcher working in the technology area of batteries. The chapter shall contribute to the advancement of more energy-efficient and environmentally friendly energy storage solutions through a comprehensive analysis of the advantages, challenges, and prospects of RRP. Schematic representation of redox-active pathway for energy storage system shown in Fig. 1.

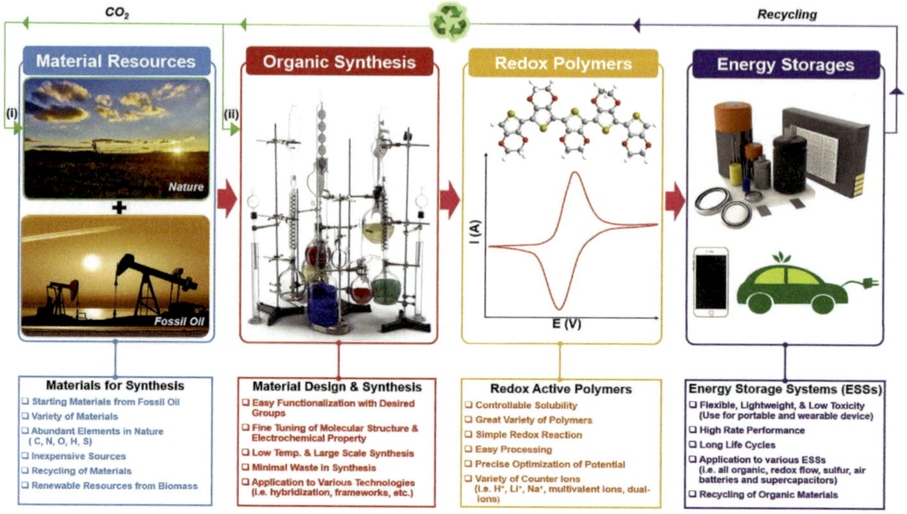

Fig. 1. Schematic representation of the synthetic pathway of redox-active polymers and their use in energy storage systems. From the above figure, it can be analyzed that organic synthesis is linked to renewable energy concepts as far as interweaving green chemistry practices into developing redox-active polymers for efficient energy storage in the forms of rechargeable batteries and supercapacitors (Copyrights from Redox-Active Polymers for Energy Storage Nano architectonics (Kim et al.,), 2017, Joule).

2. Redox Polymer Synthesis

Redox polymers were developed in the early 1980s as a new class of electroactive materials. The initial designs included polymers containing ferrocene or utilizing conducting backbones such as polyacetylene. Today, redox polymers encompass a wide range of chemical structures that combine various conjugated polymer backbones with electroactive organic, inorganic, or organometallic components. Traditionally, redox polymers have been classified based on three main criteria: (i) the positioning of the redox centre, whether it is along the polymer backbone or as a pendant group; (ii) the nature of the polymer backbone, distinguishing between conjugated/semi-conducting and non-conjugated; and (iii) the type of redox centre, which can be either organic or inorganic (Shirota et al., 1984). Although many classifications exist for synthesized and reported redox polymers, these classifications often allow for exceptions and hybrid forms (Fig. 2).

2.1 Polymers with Related Substituted Compounds

2.1.1 Ferrocene Substituted Polyethylene

Polymer films exposed to dichloromethane (CH_2Cl_2) solutions contained roughly 2 M of redox-active sites (Hunter et al., 1987). Poly vinyl ferrocene (PVFC) dissolved in CH_2Cl_2 can be potentiostatically oxidized in the presence of ferricenium salts and mixed-valence substituents to form an insoluble, doped polymer. Depending on the conditions, the resulting polymer films showed doping levels between 0.59 and

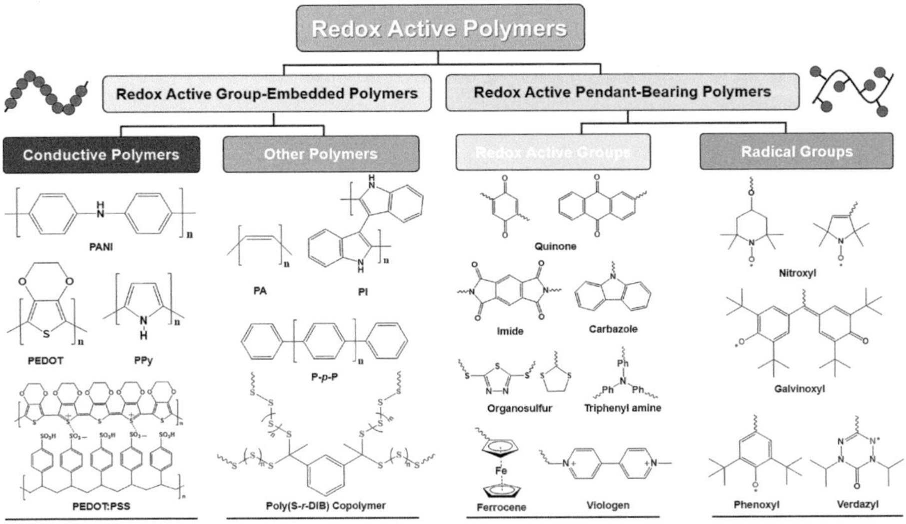

Fig. 2. Classification of redox-active polymers into different categories based on their chemical structure and application with several key redox-active functional groups commonly found in these polymers, such as quinone, pyrrole, and ferrocene groups, highlighting their role in enabling electron transfer and redox reactions (Copyrights from Redox-Active Polymers for Energy Storage Nano architectonics (Kim et al.,), 2017, Joule).

0.91, with the degree of doping showing an inverse correlation with their modest conductivities (Shirota et al., 1984).

2.1.2 Carbazole Substituted Polyethylene

In dry non-aqueous solutions, the polymer-substituted material is exposed to controlled potential electrolysis. In this case, PVCz, a type of synthetic non-aqueous polyethylene, is polymerized using the pendant vinyl group of n-vinyl carbazole. Although the doped polymer is developed rapidly, the conductivity level of the product in this synthesis is very low. Due to the selective linkage of pendant carbazole groups, the oxidized (doped) material exhibits low solubility (Jeon et al., 1992). It was recently realized that rotating electrodes can be used for potentiostatic growth control of PVCz films with a result of dense strongly adhering films. Also, mixtures of toluene-cyclohexanone can be used to prepare electrodes coated by casting films from real polymer solutions of PVCz (Compton et al., 1986; Skompska and Peter, 1995).

2.1.3 Polyoxyphenazine

In strongly acidic aqueous solutions, there is potential for the electrochemical conversion of 1-hydroxy phenazine (1-phenazinol) into polyoxyphenanthione or polyphenalzoate, known as poly (1-hydroxy phenazine) (Haas and Zumbrunnen, 1981). Polymerization results in an ether bridge between monomers that then get bridged in the chain. While the presence of an IR band at 1220 cm^{-1} indicates the possibility of an oxygen lone pair and aromatic sp^2 electrons mixing to an extent, the redox centres themselves are inherently isolated. Solutions containing $FeSO_4$ also can be used to prepare the POPh as a brown powder, this time through chemical persulfate oxidation (Miras et al., 1992).

2.1.4 Tetrathiafulvalene Substituted Polystyrene

To create electroactive modified electrodes, the polymer was coated on the substrates through a spin coating from a solution of THF. Films of different thicknesses, positioning the TTF groups 8 Å apart, yield these redox groups with a molar concentration of 3.5 M. Electron contact between these closely spaced TTF units explains how electrical conduction so readily occurs across amorphous films. The flow of counterions through the polymer matrix coincides with the passage of electrons along these groups (Blanchard et al., 1998).

2.1.5 Polymer Adducts

Adducts can be formed between some nonconducting polymers and electroactive molecules like I_2. They can be used as positive electrode materials in rechargeable batteries when mixed with a conductive filler. The current collector was a carbon fibre plate, and a rechargeable electrode was prepared by mixing nylon-6 (a polyamide, powder, or solution in formic acid) and carbon black in a 5:1 ratio. Similarly, a 5:1 polymer-carbon composite was prepared using the linear polymer PTHF that is chemically generated through chain opening from THF; however, it is generally known as a polyalkylene oxide rather than as a poly-THF (Yamamoto et al., 1984).

2.1.6 Polymer Complex

These complexes are prepared by functionalizing PVP with ruthenium ions complexed with bipyridine. The material can be spun onto GC, Au, Pt, etc., as a thin film. When the pendant pyridine group in PVP acts as a ligand to the Ru ion, it bonds with the Ru complex (Haas and Vos, 1980). The study shows that the ruthenium compounds are bonded to about one in every 4–20 PVP monomer units.

2.1.7 Redox Polymer as Electrode Material

The use of redox polymers as materials for battery electrodes has been reviewed multiple times. In the following sections, we will discuss a few selected examples in greater detail, focusing on important battery parameters such as charge/discharge potential and specific capacity. A rechargeable battery utilizing redox-active polymers as cathode materials can be paired with either a metal or redox-active polymer anode. The configuration depends on the type of redox-active groups present as shown in (Fig. 3).

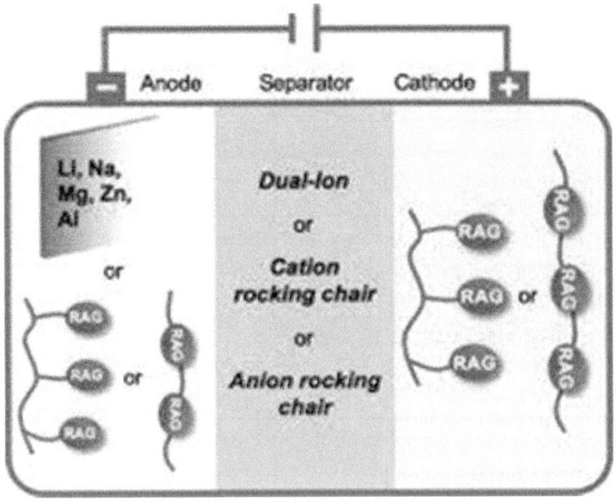

Fig. 3. Schematic diagram of a rechargeable battery featuring redox-active polymers as cathode materials, paired with a metal or redox-active polymer anode. The charge balancing is achieved by moving anions or cations from the electrolyte, depending on whether the redox-active groups (RAGs) are p-type or n-type. (Copyrights from Redox Polymers as Electrode-Active Materials for Batteries (Birgit et al.,), 2019, Organic Materials).

3. Radical Molecules

A special class of organic molecules containing one or more unpaired electrons is called radicals. Because they have an open-shell electronic configuration, radicals are generally unstable and therefore highly reactive. They often act as intermediates in chemical reactions, especially photochemical reactions, where the light energy is the initiating agent of the reaction, and in combustion reactions, where rapid oxidation is involved. In polymer chemistry, radicals play a critical role as initiators in olefin polymerization. Radicals break the double bonds of olefins so that the molecules can combine and form large polymer chains. These reactions can be terminated by radicals and can thus control the molecular weight of the resulting polymers. Though organic radical molecules are intrinsically unstable, they can be converted into stable compounds (Rozantsev and Sholle, 1971). This transformation is significant for applications, as the more stable derivatives can last longer, allowing them to be more amenable to manipulation in the open atmosphere. One powerful approach toward achieving this stability is the inclusion of sterically protected groups, where the latter may protect reactive sites on the radicals themselves, or π-conjugated chemical structures, which provide delocalization of the unpaired electrons. These modifications significantly enhance the safety and usability of radicals in various chemical applications (Table 1).

The exploration of stable free radicals has revealed some remarkable examples, in particular, relating to nitroxide radicals such as TEMPO (2,2,6,6-tetramethylpiperidine-1-oxyl), nitrogen- and oxygen-centered radicals such as 1,1-diphenyl-2-picrylhydrazyl (DPPH), and the 2,4,6-tri-t-butylphenoxy radical, and some other π-conjugated radicals such as the tris(pentachlorophenol)

Table 1. Summary of key polymers and electrochemical properties of various polymer electrode materials.

Types	Materials	Advantages	Disadvantages
Radical polymer	Phenoxyl, nitroxide, verdazyl polymers	High voltage	Slow reaction kinetics, low conductivity, low capacity
Conjugated polymers	Covalent organic frameworks, Graphdiyne, Conjugated microporous polymers	Fast reaction kinetics, high porosity, long cycle life	Low conductivity, low density, low capacity
Carbonyl polymer	Polyimides, quinone polymers	Stable voltage output, fast reaction kinetics, high capacity	Low voltage cathodes, low conductivity
Polymer-based C-N bonds	Pyrazine-based polymers, Schiff bases	Fast reaction kinetics, high capacity	Low conductivity
Conducting polymers	PANI, PAc, PPy, PTh	High conductivity	Low capacity, poor stability
Polymers of arylamines	Carbazole, triphenylamine, phenazine, phenothiazine	High voltage	Low capacity
Organosulfur polymers	Thioether, disulphides, polysulfides	High Capacity	Low conductivity, low cycle life, high solubility, low voltage.

methyl radical. Within the nitroxide species, TEMPO is uniquely characterized by its impressive reversibility in redox cycles (DaRooge and Mahoney, 1967). More specifically, TEMPO can be oxidized to its oxoammonium cation, which is a positively charged species and can be reduced back into its original radical form. This redox process involves an electron self-exchange reaction that proceeds with an extremely high-rate constant of $10^9\ M^{-1}\ s^{-1}$. This is high enough to be the indication of a fast reaction; thus, TEMPO serves as a versatile molecule in chemistry and material sciences for many applications. TEMPO's oxidized version and its salts in various counter anions including PF_6^- can also be isolated as crystals. Studies of structural properties depend, among other things on the crystal form of both TEMPO as well as its oxidised analogue. Importantly, the two crystal structures present very little structural variation as a result of this one-electron transfer reaction. This further supports the simplicity and speed of this redox reaction. Also, the fact that it was possible to crystallize the TEMPO cation salt points to the chemical stability of both the starting TEMPO radical and its oxidized cation. This stability leads to a phenomenon called "bistability," in which both states coexist with a small difference in Gibbs free energy and low activation energy barriers. These conditions are supposed to enable a super-fast and efficient electron exchange reaction between the reduced and the oxidized states. This can also be attributed to the enhanced density of redox species found in the crystalline forms of TEMPO and the cation, which makes a good case for their widespread usage in organic electronics and catalysis, and many more in energy storage technologies (Ferreri and Chatgilialoglu, 2012; Hicks, 2011;

Trapp et al., 1966). The remarkable stability and bistability of TEMPO and its related radicals position them as key solutions to critical challenges in radical chemistry and materials science.

3.1 Electrode-active Materials

Equipment that use electro-active materials include portable telephones, personal computers, robots, and electrical cars. Their operation is achieved using a rechargeable source of energy. At the centre of portable electrical equipment lies a recharged battery, a piece of modern technological engineering, which provides a means of energy that may subsequently be converted to electricity. A recharged battery essentially has two electrodes, an anode and a cathode. The two electrodes are suspended in an electrolyte solution that allows ions to pass between them (Fig. 4). The intervening film is porous so that direct contact between the electrodes does not occur. This is critical for the efficiency and performance of the battery. During discharge, the anode undergoes a form of oxidative chemical reaction. This process involves the loss of electrons from the active material in the anode to an external circuit to power the device, and simultaneously, in that same circuit, a reduction reaction occurs in the cathode, where the electrons are gained. Through these simultaneous reactions, electric currents flows into the device to energize it. In this way, the charging reverses the flow of the electrons. The moment a battery is connected to a power source, the electrons are pushed back into the anode. This triggers opposite reactions at both electrodes, where the anode recovers electrons and the cathode lets them return to the external circuit. The materials at both electrodes are involved in reversible redox reactions, implying that they can also store and release energy efficiently. The recharging of the battery by reversing the chemical reactions by several cycles makes it useful. Secondly, the ions moving around in

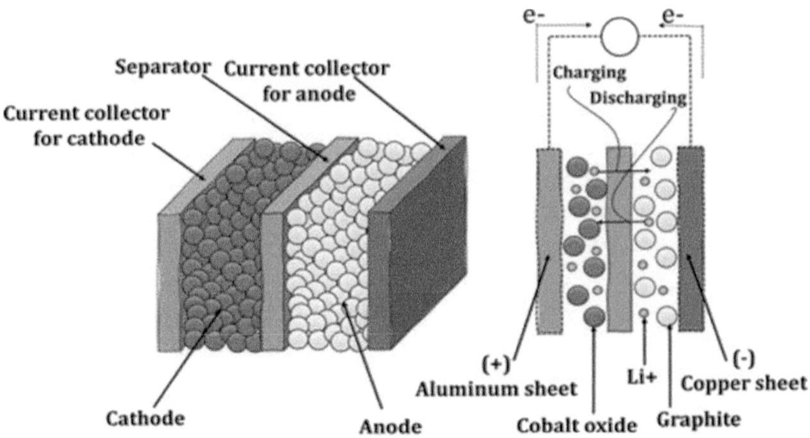

Fig. 4. Schematic representation of the cell configuration of a rechargeable battery (Copyrights from 'A Review on Composite Phase Change Materials and Fins-Based Li-Ion Battery Thermal Management Systems with Design Perspectives and Future Outlooks', Md Golam et al., 2024, ACS Energy and Fuels.)

the electrolyte solution achieve charge balance in the battery so that discharging and charging are done in stable conditions. In general, the design and functionality of rechargeable batteries are integral to the use of portable electric devices as they store energy chemically and convert it into electrical energy when needed.

The conventional rechargeable battery includes metals and metal oxides as active materials, which comprise zinc, lead, cadmium, manganese, nickel, and cobalt oxides (Reddy and Linden, 2011). These batteries possess energy density high enough, stable output voltage, and low self-discharge rates. Although mature, yet there are unsorted problems like suboptimal power performance and 41-hr charging times. Insertion and de-insertion of lithium ions from the cobalt oxide lattice within the cathode is a critical process in the battery while charging and discharging. Other limitations in using and storing metal-based batteries include toxicity, environmental hazards, overheating, complex recycling, and scarcity of metal resources (Meyers, 2012; Tarascon and Armand, 2001). Organic compounds, including conjugated carbonyl compounds and conjugated carboxylates attracted attention as good candidate materials for organic electrodes; rhodizonic and terephthalic acids are examples. Their reversible reduction may appear to involve multiple electron transfers, from the lithiation of the phenolic alcohols, and aromatic carboxylic acids. From the strongly compact molecular structure, an enormous theoretical capacity of 200–600 Ah kg^{-1} is calculated. However, the introduction of lithium ions into organic crystals and the slow lithiation reactions greatly limit the power rate performance of these electrodes, just like for the redox reactions happening in inorganic electrode active materials. Moreover, the output voltage of these organic compounds concerning Li/Li^+ is still at a very low level, less than 2.0 V, due to their negative redox potentials. Polycyclic π-conjugated molecules, for example, trioxotriangulene, are expected to exhibit a theoretical redox capacity of more than 200 Ah kg^{-1} via multiple redox reactions. However, these multiple reactions yield low power rates, output voltages, and, in many cases, lead to chemical instability of the molecules. The organic compounds used as electrode-active materials were normally tested in their crystalline or solid powder forms that lack electrical conductivity. This requires a high content of conductive carbon that significantly reduces the electrode's practical energy capacity (Armand et al., 2009; Jouhara et al., 2018; Poizot et al., 2018).

Tetraethyl thiuram disulfide, along with other compounds in the family of diphenyl disulfides, have gained special attraction recently due to the ability they may exhibit in their structures for being used as a cathode-active material for a battery. With voltages between 2.0 and 2.3 V regarding the Li/Li^+ standard, it gains great importance to become a suitable material for the fabrication of a battery. Being calculated with an astonishing theoretical capacity value of 4800 Ah kg^{-1} these materials would add significantly to achieving high energy densities with their batteries. However, the main disadvantage is the slow kinetics in the redox reaction of the disulfides with thiols, or vice versa, their lithium thiolates. The reported rate constant for the electron-transfer reaction is as small as 10^{-8} cm s^{-1}, which is a significant hindrance to the overall power rate performance of batteries using such materials. The problem of slow electron transfer is under study using several different approaches. The sulfur molecule can be modified for the enhancement of its reactivity; conductive additives are incorporated into the material to enhance electron flow through the material. The compounds continue to face significant difficulties due to poor structural organization and the low order of the sulfur fragments (Chung et al., 2013; Visco et al., 1989). The nitroxide radical species, including TEMPO [2,2,6,6-tetramethylpiperidine-N-oxyl], emerge as a second promising candidate class of redox-active materials. TEMPO showed electrochemically reversible redox behavior at relatively high potentials of up to approximately 3.7 V in a rechargeable beaker cell configuration using $LiPF_6$ as an electrolyte. The capacity of TEMPO is itself less compared with organosulfur compounds but, due to its high operating voltage and unique electrochemical properties, it becomes very attractive. Importantly, the redox reaction of TEMPO does not interfere with ambient oxygen due to its positive position of potential. This further makes it suitable for use in batteries, especially under oxidative conditions that might be harmful to other materials. The redox reaction of TEMPO is compatible with current collectors that are conventionally used, such as glassy carbon plates, ITO films, and aluminium foil. One of its advantages is the remarkably high heterogeneous electron-transfer rate constant at around 10^{-1} cm s^{-1}. This rapid electron exchange reaction contributes to its very superior power rate performance with TEMPO used as an active material in the cathode. The favourable redox properties of TEMPO together with the stability of its cationic form offer a quasi-solidification property while keeping counterion mobility, which is very useful in the construction of high-performance rechargeable batteries (Nakahara et al., 2011a; Nishide and Suga, 2005). Besides, it has been found that TEMPO analogs in polymer form are quite feasible and have widened their applications to various configurations of batteries and could thus bring new energy storage devices into the future.

3.2 Synthesis of Radical Polymers

TEMPO polymers have garnered significant attention in recent research due to their potential as environmentally friendly, metal-free catalysts in the oxidative reactions of various organic molecules. These reactions include the conversion of alcohols into aldehydes and carboxylic acids, which are important processes in organic synthesis. The unique structure of TEMPO polymers is characterized by cross-linking, which allows them to function as heterogeneous catalysts. This means they can facilitate chemical reactions in a solution while remaining distinct from the reactants, enabling easy separation through simple filtration methods, similar to the operations of an ion-exchange resin. Despite their advantages in catalytic applications, TEMPO polymers have not been widely explored as electrode-active materials. In a novel approach, the authors of the study shifted their focus to evaluating the potential of redox-active organic compound polymers (Endo et al., 1993). They examined these polymers by analysing their redox potential and overall storage capacity, which are critical factors for their performance in electrochemical applications. Accordingly, radical molecules such as TEMPO are regarded as 'radical polymers,' distinguished by their redox reactivity and suitability for electrode-active applications. The identification and utilization of radical polymers could pave the way for innovative developments in energy storage and conversion technologies, thereby expanding the scope of applications for TEMPO and similar polymers in the field of electrochemistry. Electrode active materials include:

i. Solvophilic or well-solvation with an electrolyte is necessary for radical polymers to function as electrode-active materials and to guarantee counterion movement within the polymer.
ii. A sufficiently inflated yet insoluble material in the electrolyte solution to prevent self-discharging. The disintegration of the polymer or the elution of the redox-active component in the electrolyte solution acts as a charge-carrier transfer between the cathode and anode, resulting in the lethal phenomena of self-discharge. Radical polymers are immobilized in the manufactured electrode due to the creation of exceptionally high-molecular-weight polymers and/or minor cross-linking of polymer chains.
iii. The tightly packed structure of the radical site's repeating or monomer unit. The generated polymer backbone sacrifices the energy density or redox capacity of the radical polymers.
iv. In a radical polymer, charge propagation or electron exchange is frequently facilitated by the pliable polymer framework and its linker group that binds the backbone to the radical functional group.

The redox activity of the TEMPO (2,2,6,6-tetramethylpiperidin-1-yl) moiety plays a significant role in hindering conventional radical polymerization processes. It disrupts the interactions between initiating and propagating radicals, ultimately preventing the formation of the desired polymers when using typical vinyl monomers. However, there is a viable alternative in the form of anionic polymerization of TEMPO-containing vinyl monomers, which can yield highly accurate and high-molecular-weight radical polymers. Conducting the reactions in well-dried glass vessels is crucial to effectively utilize this anionic polymerization approach, as moisture can adversely affect the reaction conditions. Skilful handling and experimental techniques are also necessary to achieve satisfactory results. One specific example of this process is the anionic polymerization of TEMPO methacrylate, a type of vinyl monomer. This polymerization is initiated using methyl methacrylate-capped diphenyl hexyl lithium, which is the initiator. The chosen initiator presents a moderate degree of nucleophilicity. This characteristic is essential as it helps prevent the nitroxide radical associated with the TEMPO moiety from engaging in reactions with the carbanion generated by diphenyl hexyl lithium. As a result, the process enables the production of poly (TEMPO methacrylate) with several desirable attributes: it features a well-defined molecular weight, a narrow polydispersity index, a high yield, and maintains an accurate ratio of 1.0 radical per monomer unit. In addition to the anionic approach, cationic polymerization is another method for synthesizing TEMPO-bearing vinyl polymers (Qu et al., 2007; Suguro et al., 2007). In this case, the starting materials include a TEMPO-substituted vinyl monomer, which is characterized by an electron-donating ether linkage. The cationic polymerization process can be initiated using boron trifluoride etherate. This method successfully yields the corresponding TEMPO polymer, which possesses a high theoretical capacity. Furthermore, this polymer offers considerable flexibility and lipophilicity, making it versatile for different applications. An added advantage of this approach is that it minimizes any bond cleavage reactions, such as hydrolysis, which could otherwise compromise the integrity of the resulting polymer.

3.3 Working Principle

Rechargeable batteries can incorporate polymers as electrode materials, specifically when these polymers employ stable organic radicals as electrochemically active pendant groups. This innovative approach delivers outstanding power densities, rapid recharge times, and remarkable cycling stability, making it a notable advancement in battery technology. The use of the persistent radical 2,2,6,6-tetramethylpiperidinyloxyl (TEMPO) marked a significant milestone in battery research; it was first reported by Nakahara et al. in 2002 (Nakahara et al., 2002). In their study, poly(methacrylate) was utilized as the active cathode material, known as PTMA. Over the years, this system has been documented in numerous original research articles and comprehensive review papers, evolving into a widely studied standard model, especially when paired with various lithium anode configurations. This combination has effectively established a new category of batteries known as organic radical batteries. In these batteries, the active electrode material is a stable polymer that incorporates nitroxide radicals, which play a crucial role in the battery's electrochemical processes. Batteries that are classified as lithium organic, or all organic radical batteries, depend significantly on the type of counter electrode used. The nitroxide radicals can participate in two distinct reversible reactions: the first is the reduction of the radical to generate an amino acid anion, and the second is the oxidation of the radical that results in the formation of an oxoammonium cation (Kim et al., 2007; Wang et al., 2011). The first reaction contributes to polymer doping through reduction, while the second reaction facilitates polymer doping through oxidation. Therefore, an ideal organic radical battery incorporates two different nitroxide radicals. One radical is selected for its redox potential to serve as the anode, while the other functions as the cathode. During charging, the nitroxide radical gets reduced at the cathode, forming an aminoxyl anion while oxidation at the anode it forms an oxoammonium cation. When the battery is discharged, the process is reversed, and the related nitroxide radicals are regenerated. Higher energy densities and cell voltages can be achieved by combining a radical polymer cathode with lithium metal or lithium insertion materials. This strategic pairing within the lithium organic battery system not only improves efficiency but also greatly advances battery technology, paving the path for more effective and sustainable energy storage options (Fig. 5).

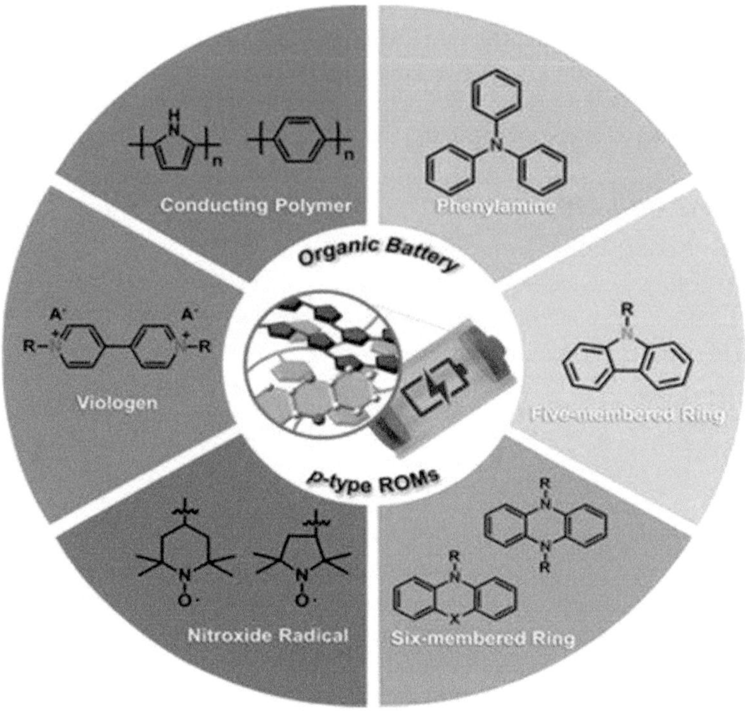

Fig. 5. Design and mechanism of rechargeable batteries with redox-active organic radicals (Copyrights from 'p-Type Redox-Active Organic Electrode Materials for Next-Generation Rechargeable Batteries' (Hyojin et al.,), 2022, Advanced Energy and Sustainability Research).

4. Types of Radicals and Reactions

4.1 NO Radicals

NO radicals, also known as aminoxyl radicals, have the following structure,

$$R_2N{-}O\bullet \leftrightarrow R_2N\bullet + {-}O{-}$$

Although IUPAC does not recommend the term "nitroxide," it remains widely used among professionals. Electron spin resonance (ESR) has been instrumental in examining the spin properties of early nitroxides like di-tert-butyl nitroxide and Fremy's salt. In these compounds, bulky methyl or sulfonyl groups protect the nitroxide functionalities from detrimental radical side reactions. Nitroxides with strong NO radicals commonly display p-type redox activity, wherein oxoammonium cations form from their neutral radical states. In the 1960s, researchers enhanced the stability of nitroxide radicals by restricting bond rotation in aliphatic cyclic structures, leading to potent radicals such as TEMPO, 2-azaadamantane N-oxyl (AZADO), and 2,2,5,5-tetramethyl pyrrolidine 1-oxyl (PROXYL) (Oyaizu et al., 2010). When n-type redox states are accessible, reduction results in the formation of aminoxyl anions. By examining various molecular structures, scientists have fine-tuned the redox properties of nitroxides. This review explores the redox behaviours of several nitroxides, including TEMPO, AZADO, and PROXYL. With its relatively lower molar mass compared to the TEMPO radical, PROXYL offers a higher specific capacity than other compounds with similar functionalities in the five-membered ring structure of 2,2,5,5-tetramethyl-2,5-dihydro-1H-pyrrol-N-oxyl (NPhP). PROXYL-based units in a poly (ethylene oxide) (PEO) backbone have shown a cell voltage of 3.6 V, which is higher than that observed with TEMPO-containing polymers, such as PTGE, in lithium-organic batteries. Literature indicates that a theoretically achievable capacity of 147 Ah kg^{-1} could be obtained using polymers with PEO and PROXYL groups as nitroxyl-radical bearers. In practical tests, the battery achieved less than half of this theoretical capacity; however, it remained stable, retaining 80% capacity over 1,000 cycles, and reached a cell voltage of 7 V. The nitronyl nitroxide radical, as a redox-active polymer, can be both oxidized and reduced, making it a viable candidate for both anode and cathode applications (Nesvadba et al., 2010). The polymerization process yielded a functionalized poly(acetylene) incorporating nitronyl and azoxide groups, which displayed reversible redox behaviour at 0.80 V and −0.84 V versus AgCl/Ag. While this compound has not yet been implemented in batteries, electrochemical studies highlight its potential for battery applications.

4.1.1 Galvinoxyl

The nitroxide-free radical galvinoxyl can, in a reversible coupling reaction, convert into its phenolate form. This is relevant because, though the redox potential of galvinoxyl is rather high, this radical serves as an efficient n-type material in numerous all-organic battery systems. Galvinoxyl is a member of the quinones series and, together with other quinones having single-electron redox potential, leads to their great molar mass for having a theoretical reaction capacity as high as 51 Ah kg^{-1}, hence galvinoxyl could be a rather attractive compound in terms of energy storage. A particular ORB based on poly (ethynylphenyl galvinoxyl) anode, PTMA cathode, and aqueous electrolyte medium was demonstrated by constructing in the experiments. Yet, this configuration revealed a Coulombic efficiency of only 72%, a factor attributed to the performance limitations of the PTMA cathode (Suga et al., 2011). In this system, the charge and discharge plateaus were only faintly defined, indicating that the efficiency of energy transfer was not optimal. Operational conditions also play a crucial role in performance outcomes.

PTMA operates best at near-neutral to slightly acidic pH conditions, whereas the galvinoxyl polymer works best in an alkaline condition. This would suggest that the neutral aqueous NaCl solution used in the original battery assembly was not in the best condition for optimal operation. Alternative configurations, in which a poly (nitronyl nitroxyl styrene) anode and a poly (galvinoxyl styrene) cathode were employed, also showed promising performance. The cell had a storage capacity of 29 Ah kg^{-1} at an output voltage of 0.6 V and was thus at 91% of its theoretical maximum. In this assembly, the nitronyl nitroxide is reduced in the charging process, allowing the cations to be taken into the system. At the same time, the reduced galvinoxyl reoxidizes and liberates an equivalent number of counterions. As a result, both half-cells in this configuration only host neutral and negatively charged states. The interaction between these states allows for the exchange of countercations between the electrolyte and the polymer matrices, facilitating a "rocking-chair" migration mechanism. This mechanism contributes to the reduced requirement for supporting electrolytes, as the unbound counterions concentration can remain stable, independent of the battery's charge state. Additionally, researchers are looking into other potential radical compounds such as Blatter's radical, oxyl-based phenoxyl radical, and verdazyl. Although these compounds have yet to be implemented in organic battery systems, they represent exciting avenues for future research. While radical polymers account for a substantial percentage of published literature in the organic battery field, the variety of radicals remains relatively limited when compared to the vast array of employed organic compounds (Nakahara et al., 2011b). Nonetheless, it is important to note that radicals exhibit exceptional electrochemical properties, particularly in terms of their redox kinetics. Among these radicals, TEMPO stands out as the most extensively examined moiety for polymer-based batteries, underscoring its considerable electrochemical stability and potential for further applications in energy storage technologies.

4.2 Applications in Energy Storage and Energy Conversion

This century is experiencing population growth and rapid technological advancements, making energy one of the most pressing issues. To meet these new demands, it has become essential to develop sustainable and environmentally friendly energy technologies. Society is increasingly shifting toward innovative, clean solutions to harness electricity from abundant natural resources such as the sun, waves, and wind. Additionally, there is now a stronger emphasis on techniques for efficiently storing the energy captured from these sources. Redox polymers play a vital role in advancing some of these technologies, which this section will highlight. Among the most prominent applications of redox polymers at present is in electrochemical energy storage devices, such as batteries, and supercapacitors, which will be discussed in this chapter (Fig. 6). While some of these technologies are still relatively immature, they are expected to grow quickly. Enzymes serve as catalysts in biofuel cells converting natural sugars into electricity, functioning similarly to the enzyme driven reactions central to electrochemical biosensor performance; redox polymers have been used in them, and they are used in thermoelectric and solar cells. It does not cover converting heat or light into electricity devices because the polymers concerned with those applications exhibit redox-like behavior and are primarily classified as (semi)conducting, semi-conjugated, or optoelectronic (Ohayon et al., 2020).

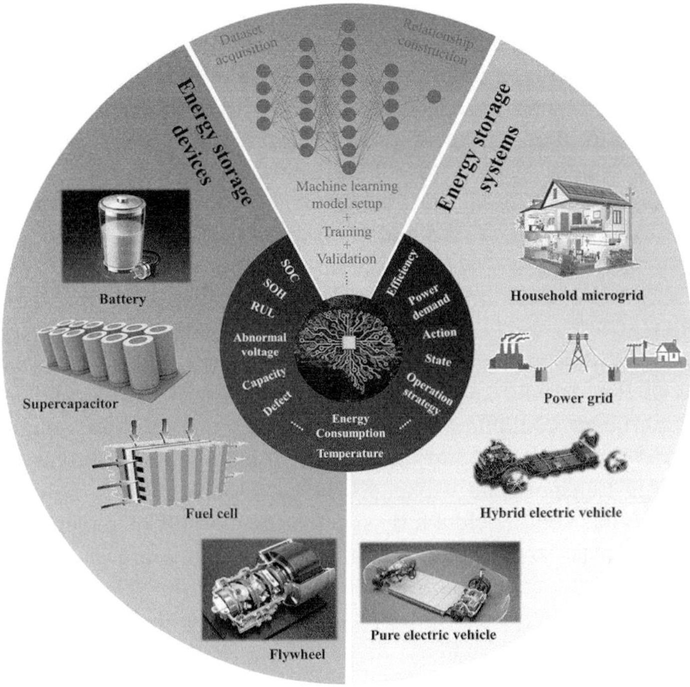

Fig. 6. Key applications of redox polymers in the energy field, include roles in energy storage devices and systems (Gao and Lu, 2021).

4.3 Energy Storage Solutions

Radical-based redox polymers are currently under active investigation for their potential in advanced electrochemical energy storage technologies, especially in various battery types. Interest in conducting polymers for battery applications first emerged in the 1980s, driven by the promise of innovative energy storage solutions. However, the initial excitement waned as stability issues plagued batteries and supercapacitors using these polymers, limiting their commercial viability and stalling progress in the field. In recent years, a renewed focus on redox polymers has brought them back into the spotlight within several emerging battery technologies. Notable applications include flexible thin batteries, offering lightweight and adaptable energy storage solutions; fast-charging batteries, designed to drastically reduce recharge times; organic batteries, which utilize sustainable materials; metal-air batteries, known for high energy densities; redox-flow batteries, which allow for scalable and efficient energy management; and aqueous batteries that emphasize safety and environmental sustainability. In these applications, redox polymers can function in a very versatile set of roles as active materials in the cathode or anode; as redox-active binders or mediators to improve charge transfer at the interfaces; or in aqueous redox flow battery systems as soluble catholytes or anolytes. This text scopes out the important redox polymers for battery technologies-including specific challenges, advantages, and essential characteristics. It provides an in-depth analysis of how different polymer families such as radical polymers, capable of rapid electron transfer; phenothiazine-type polymers, valued for their stability and environmental compatibility; carbonyl polymers, which offer efficient redox activity; and catechol polymers, known for their strong interactions with metal ions are being incorporated into innovative battery designs to enhance performance. In the field of organic radical batteries (ORBs), molar mass contributes to the limitation of organic radical polymers' capacity because the amount of charge it can store is directly proportional to its relative molar mass. These are, however, somewhat alleviated by polymers with promising redox potentials compared to traditional metal oxides like lithium cobalt oxide ($LiCoO_2$) for TEMPO-containing poly(2,2,6,6-tetramethylpiperidinyl) (PTMA). This implies that they may replace conventional materials and perhaps reduce one of the drawbacks of battery's production and lifecycle dispositions. However, significant challenges remain, including the need to reduce both the molar mass per repeating unit and the number of conductive additives required for effective charge transport and capacity. The development of a high-performance all-organic ORB has yet to yield a fully successful system. Current proof-of-concept implementations face capacity limitations due to an unfavourable charge-to-molar-mass ratio for each repeating unit. Recent innovations, however, have introduced novel approaches. Poly(nitronylnitroxylstyrene), a bipolar redox-active radical polymer, has enabled new battery architectures, including a pole-less battery configuration and a rocking-chair-type battery paired with poly(galvinoxylstyrene). One promising advance is a flexible metal-free battery with a discharge plateau around 0.7 V, showcasing its potential for practical applications. Additionally, an all-organic battery was created using poly (TEMPO-vinyl ether) as the cathode and an anthraquinone-substituted poly(styrene) as the anode, marking another step forward in organic battery technology. The most promising approach to date involves pairing a TEMPO-substituted poly(acrylamide) cathode with poly(viologen) as the anode. This configuration has achieved a competitive cell voltage of around 1.2 V and demonstrated excellent cycling performance, suggesting a promising path for developing efficient and sustainable organic battery systems (Tang et al., 2018). The research and insights presented in this text aim to highlight the transformative potential of redox polymers in the future of battery technology.

5. Conclusion

The study of redox-active organic compounds, especially in polymer form, has been a dynamically evolving field in secondary battery applications over the past several decades. This review aims to provide an in-depth analysis of polymer-based organic batteries on the diverse range of redox-active materials and their critical characteristics, operational mechanisms, and essential non-active components, such as conductive additives, current collectors, and electrolytes, which work together to enhance battery performance. This review introduces characterization techniques to scrutinize such innovative battery systems, thereby commencing with a set of methodologies that, first of all, measure not only performance metrics but start with untangling the intricate electrochemical processes into these systems. This work touches on the socio-economic implications of using such state-of-the-art energy storage systems for sustainable energy solutions in contrast to relying on traditional materials which pose an environmental and ethical burden in most cases. Such redox-active compounds can adopt a systematic classification based on the identity of their electrochemically active functional groups. This will provide a comprehensive study of redox mechanisms and electrochemical properties across categories. Such a section would provide relevant information about significant differences in key performance metrics, including specific capacity, cycling stability, and voltage characteristics. Understanding such distinctions is important for determining which material is more practically viable than others in real applications with batteries. They come out as quite promising contenders for being considered as batteries' electrodes. These turn out to be TEMPO-based polymers as cathode while anthraquinone-based compounds as anode; this brings about the enhanced electrochemical properties that further include outstanding cycling stability besides general robustness in conjunction with considerable and improved energy power density. The rich structural diversity of redox-active organic compounds allows for fine-tuning of the cell potential, making them capable of being tailored to any energy storage requirement. Based on a diverse set of chemical building blocks derived largely from renewable resources, scientists are now in a good position to take the first steps toward creating advanced organic materials with vast promise for next-generation rechargeable electrochemical energy storage devices. This area is still in its early stages as commercial systems for applying redox-active polymers to secondary batteries are yet to materialize. However, the field offers scope for innovation. Perhaps the conversion of battery systems to sustainable and renewable materials will prove to be an important move toward the realization of rising global demands for efficient and environment-friendly energy storage alternatives, ultimately leading toward a more sustainable energy future.

References

Akinyele, D.O. and Rayudu, R.K. 2014. Review of energy storage technologies for sustainable power networks. Sustainable Energy Technologies and Assessments 8: 74–91.

Andersson, B.A. and Råde, I. 2001. Metal resource constraints for electric-vehicle batteries. Transp Res. D Transp. Environ. 6: 297–324.

Armand, M., Grugeon, S., Vezin, H., Laruelle, S., Ribière, P., Poizot, P. et al. 2009. Conjugated dicarboxylate anodes for Li-ion batteries. Nat. Mater 8: 120–125.

Blanchard, P., Svenstrup, N., Rault-Berthelot, J., Riou, A. and Becher, J. 1998. Synthesis and characterisation of macrobicyclic tetrathiafulvalene-bridged cage molecules. European J. Org. Chem. 1743–1757.

Brandon, N.P., Skinner, S. and Steele, B.C.H. 2003. Recent advances in materials for fuel cells. Annu. Rev. Mater. Res. 33: 183–213.

Chen, C.H., Vaughey, J.T., Jansen, A.N., Dees, D.W., Kahaian, A.J., Goacher, T. et al. 2001. Studies of Mg-substituted Li4− xMg x Ti_5O_{12} spinel electrodes ($0 \leq x \leq 1$) for lithium batteries. J. Electrochem. Soc. 148: A102.

Chen, H., Armand, M., Demailly, G., Dolhem, F., Poizot, P. and Tarascon, J. 2008. From biomass to a renewable LixC6O6 organic electrode for sustainable Li-ion batteries. ChemSusChem: Chemistry & Sustainability Energy & Materials 1: 348–355.

Chung, W.J., Griebel, J.J., Kim, E.T., Yoon, H., Simmonds, A.G., Ji, H.J. et al. 2013. The use of elemental sulfur as an alternative feedstock for polymeric materials. Nat. Chem. 5: 518–524.

Compton, R.G., Davis, F.J. and Grant, S.C. 1986. The anodic oxidation of poly (N-vinylcarbazole) films. J. Appl. Electrochem. 16: 239–249.

DaRooge, M.A. and Mahoney, L.R. 1967. Reaction of 2, 4, 6-tri-tert-butylphenoxy radical with unhindered phenols. J. Org. Chem. 32: 1–6.

Endo, T., Takuma, K., Takata, T. and Hirose, C. 1993. Synthesis and polymerization of 4-(glycidyloxy)-2, 2, 6, 6-tetramethylpiperidine-1-oxyl. Macromolecules 26: 3227–3229.

Ferreri, C. and Chatgilialoglu, C. 2012. Lipid isomerization. Encyclopedia of Radicals in Chemistry, Biology and Materials.

Fukao, K., Hamagata, S., Inui, T., Ito, K., Kwon, H.U., Makino, T. et al. 2007. Estimation procedure and TFP analysis of the JIP database 2006 (revised). pp. 07-E. *In*: RIETI Discussion Paper Series. Research Institute of Economy, Trade and Industry Tokyo.

Gao, T. and Lu, W. 2021. Machine learning toward advanced energy storage devices and systems. iScience 24(1): 101936.

Haas, O. and Vos, J.G. 1980. Preparation, characterisation and electrochemistry of electrodes coated with poly (4-vinylpyridine) bound ruthenium complexes. J. Electroanal. Chem. Interfacial Electrochem. 113: 139–149.

Haas, O. and Zumbrunnen, H. 1981. Electrochemical properties of hydroxyphenazine-coated electrodes. Helv Chim Acta 64: 854–863.

Hameer, S. and van Niekerk, J.L. 2015. A review of large-scale electrical energy storage. Int. J. Energy Res. 39: 1179–1195.

Häupler, B., Wild, A. and Schubert, U.S. 2015. Carbonyls: powerful organic materials for secondary batteries. Adv. Energy Mater 5: 1402034.

Hicks, R. 2011. Stable Radicals: Fundamentals and Applied Aspects of Odd-Electron Compounds. John Wiley & Sons.

Hunter, T.B., Tyler, P.S., Smyrl, W.H. and White, H.S. 1987. Impedance analysis of poly (vinylferrocene) films: the dependence of diffusional charge transport and exchange current density on polymer oxidation state. J. Electrochem. Soc. 134: 2198.

Huo, M., Yuan, J., Tao, L. and Wei, Y. 2014. Redox-responsive polymers for drug delivery: from molecular design to applications. Polym. Chem. 5: 1519–1528.

Inagaki, M., Konno, H. and Tanaike, O. 2010. Carbon materials for electrochemical capacitors. J. Power Sources 195: 7880–7903.

Janoschka, T., Hager, M.D. and Schubert, U.S. 2012. Powering up the future: radical polymers for battery applications. Advanced Materials 24: 6397–6409.

Jeon, I.-R., Noma, N., Claridge, R.F.C. and Shirota, Y. 1992. Electrochemical doping of Poly(3-vinylperylene) and electrical properties of doped polymers. Polym J. 24: 273–279. https://doi.org/10.1295/polymj.24.273.

Jouhara, A., Dupré, N., Gaillot, A.-C., Guyomard, D., Dolhem, F. and Poizot, P. 2018. Raising the redox potential in carboxyphenolate-based positive organic materials via cation substitution. Nat. Commun. 9: 4401.

Kim, J.-K., Cheruvally, G., Choi, J.-W., Ahn, J.-H., Lee, S.H., Choi, D.S. et al. 2007. Effect of radical polymer cathode thickness on the electrochemical performance of organic radical battery. Solid State Ion 178: 1546–1551.

Kim, J., Kim, J.H. and Ariga, K. 2017. Redox-active polymers for energy storage nanoarchitectonics. Joule 1: 739–768.

Linden David, R.T.B. 2002. Handbook of Batteries 3rd ed, ed. M. H. handbooks2002.

Lu, Y., Zhang, Q., Li, L., Niu, Z. and Chen, J. 2018. Design strategies toward enhancing the performance of organic electrode materials in metal-ion batteries. Chem 4: 2786–2813.

MacDiarmid, A.G. 2001. "Synthetic metals": a novel role for organic polymers (Nobel lecture). Angewandte Chemie International Edition 40: 2581–2590.

Meyers, R.A. 2012. Encyclopedia of Sustainability Science and Technology. Springer New York.

Miras, M.C., Barbero, C., Kötz, R., Haas, O. and Schmidt, V.M. 1992. Quartz crystal microbalance and probe beam deflection studies of poly (1-hydroxyphenazine) modified electrodes. Journal of Electroanalytical Chemistry 338: 279–297.

Muench, S., Wild, A., Friebe, C., Haupler, B., Janoschka, T. and Schubert, U.S. 2016. Polymer-based organic batteries. Chem. Rev. 116: 9438–9484.

Nakahara, K., Iwasa, S., Satoh, M., Morioka, Y., Iriyama, J., Suguro, M. et al. 2002. Rechargeable batteries with organic radical cathodes. Chem. Phys Lett. 359: 351–354.

Nakahara, K., Oyaizu, K. and Nishide, H. 2011a. Organic radical battery approaching practical use. Chem. Lett. 40: 222–227.

Nakahara, K., Oyaizu, K. and Nishide, H. 2011b. Organic radical battery approaching practical use. Chem. Lett. 40: 222–227.

Nesvadba, P., Bugnon, L., Maire, P. and Novák, P. 2010. Synthesis of a novel spirobisnitroxide polymer and its evaluation in an organic radical battery. Chemistry of Materials 22: 783–788.

Nishide, H. and Suga, T. 2005. Organic radical battery. Electrochem Soc Interface 14: 32.

Novák, P., Müller, K., Santhanam, K.S.V. and Haas, O. 1997. Electrochemically active polymers for rechargeable batteries. Chem. Rev. 97: 207–282.

Ohayon, D., Nikiforidis, G., Savva, A., Giugni, A., Wustoni, S., Palanisamy, T. et al. 2020. Biofuel powered glucose detection in bodily fluids with an n-type conjugated polymer. Nat. Mater 19: 456–463.

Oyaizu, K. and Nishide, H. 2009. Radical polymers for organic electronic devices: a radical departure from conjugated polymers? Advanced Materials 21: 2339–2344.

Oyaizu, K., Kawamoto, T., Suga, T. and Nishide, H. 2010. Synthesis and charge transport properties of redox-active nitroxide polyethers with large site density. Macromolecules 43: 10382–10389.

Pietschnig, R. 2016. Polymers with pendant ferrocenes. Chem. Soc. Rev. 45: 5216–5231.

Poizot, P., Dolhem, F. and Gaubicher, J. 2018. Progress in all-organic rechargeable batteries using cationic and anionic configurations: Toward low-cost and greener storage solutions? Curr. Opin. Electrochem 9: 70–80.

Qiu, X.-Y., Zhuang, Q.-C., Zhang, Q.-Q., Cao, R., Ying, P.-Z., Qiang, Y.-H. et al. 2012. Electrochemical and electronic properties of $LiCoO_2$ cathode investigated by galvanostatic cycling and EIS. Physical Chemistry Chemical Physics 14: 2617–2630.

Qu, J., Fujii, T., Katsumata, T., Suzuki, Y., Shiotsuki, M., Sanda, F. et al. 2007. Helical polyacetylenes carrying 2, 2, 6, 6-tetramethyl-1-piperidinyloxy and 2, 2, 5, 5-tetramethyl-1-pyrrolidinyloxy moieties: Their synthesis, properties, and function. J. Polym. Sci. A Polym. Chem. 45: 5431–5445.

Reddy, T.B. and Linden, D. 2011. Lithium primary batteries. Linden's Handbook of Batteries; McGraw-Hill Education: New York, NY, USA 11–14.

Rozantsev, E.G. and Sholle, V.D. 1971. Synthesis and reactions of stable nitroxyl radicals I. Synthesis. Synthesis (Stuttg) 190–202.

Shi, Y. and Yu, G. 2016. Designing hierarchically nanostructured conductive polymer gels for electrochemical energy storage and conversion. Chemistry of Materials 28: 2466–2477.

Shirota, Y., Kakuta, T. and Mikawa, H. 1984. Electrochemical oxidation of poly (vinylferrocene) with concurrent precipitation on the electrode; preparation of an electrically conducting polymer. Die Makromolekulare Chemie, Rapid Communications 5: 337–340.

Skompska, M. and Peter, L.M. 1995. Electrodeposition and electrochemical properties of poly (N-vinylcarbazole) films on platinum electrodes. Journal of Electroanalytical Chemistry 383: 43–52.

Song, Z. and Zhou, H. 2013. Towards sustainable and versatile energy storage devices: an overview of organic electrode materials. Energy Environ Sci. 6: 2280–2301.

Suga, T., Sugita, S., Ohshiro, H., Oyaizu, K. and Nishide, H. 2011. p- and n-Type bipolar redox-active radical polymer: toward totally organic polymer-based rechargeable devices with variable configuration. Advanced Materials 6: 751–754.

Suguro, M., Iwasa, S., Kusachi, Y., Morioka, Y. and Nakahara, K. 2007. Cationic polymerization of poly (vinyl ether) bearing a TEMPO radical: a new cathode-active material for organic radical batteries. Macromol. Rapid Commun. 28: 1929–1933.

Sui, X., Feng, X., Hempenius, M.A. and Vancso, G.J. 2013. Redox active gels: synthesis, structures and applications. J. Mater Chem. B 1: 1658–1672.

Tang, M., Li, H., Wang, E. and Wang, C. 2018. Carbonyl polymeric electrode materials for metal-ion batteries. Chinese Chemical Letters 29: 232–244.

Tarascon, J.-M. and Armand, M. 2001. Issues and challenges facing rechargeable lithium batteries. Nature 414: 359–367.

Trapp, C., Wang, C. and Filler, R. 1966. Electron spin resonance absorption in the Tris (pentafluorophenyl) methyl radical. J. Chem. Phys. 45: 3472–3474.

Visco, S.J., Mailhe, C.C., De Jonghe, L.C. and Armand, M.B. 1989. A novel class of organosulfur electrodes for energy storage. J. Electrochem. Soc. 136: 661.

Wang, Y.-H., Hung, M.-K., Lin, C.-H., Lin, H.-C. and Lee, J.-T. 2011. Patterned nitroxide polymer brushes for thin-film cathodes in organic radical batteries. Chemical Communications 47: 1249–1251.

Whittingham, M.S. 2004. Lithium batteries and cathode materials. Chem. Rev. 104: 4271–4302.

Xie, J. and Zhang, Q. 2016. Recent progress in rechargeable lithium batteries with organic materials as promising electrodes. J. Mater. Chem. A Mater 4: 7091–7106.

Yamamoto, T., Hishinuma, M. and Yamamoto, A. 1984. Zn | ZnI2 | iodine secondary battery using iodine-nylon-6 adduct as positive electrode, and its charge-discharge performance. Inorganica Chim Acta 86: L47–L49.

Zu, C.-X. and Li, H. 2011. Thermodynamic analysis on energy densities of batteries. Energy Environ. Sci. 4: 2614–2624.

Chapter 2

Conjugated Redox Polymers for Energy and Biomedical Applications

Atefeh Derakhshani,[1,*] *Ali KarbalaeiMahdi,*[2] *Faranak Mohamadi*[2] and *Hossein Ghanbari*[2]

1. Introduction

The discovery of polyacetylene in the late 1970s marked the beginning of significant research interest in one-dimensional (1D) conjugated polymers due to their impressive conductivity, adjustable band gap, and compatibility with solution processing, which positions them as remarkable materials for advanced organic electronics, solar cells, and thermoelectric applications (Liu et al., 2023). Conductive polymers represent a category of polymers capable of conducting electricity and play a crucial role in numerous scientific and technological applications. They are being explored as reference electrodes using different strategies. Specifically, the electrochemical alteration of the optical, conductive, magnetic, and mechanical characteristics of conductive polymers offers a promising approach to developing smart materials (Heinze et al., 2010). Redox polymers are also important materials in different energy conversion devices and technologies (Casado et al., 2016). These are macromolecules that exhibit electroactivity and have specific areas or groups capable of undergoing oxidation (electron loss) and reduction (electron gain). Conjugated polymers are a

[1] Department of Tissue Engineering, School of Advanced Technologies in Medicine, Tehran University of Medical Sciences, Tehran, Iran.
[2] Department of Medical Nanotechnology, School of Advanced Technologies in Medicine, Tehran University of Medical Sciences, Tehran, Iran.
* Corresponding author: atefe.drkhshani93@gmail.com

DOI: 10.1201/9781003546672-2

subcategory of polymers that have unique molecular backbone structures, which are composed of alternating single and double bonds. They are typically formed by polymerizing free radicals of unsaturated alkenes or heterocyclic compounds, i.e., acetylene, pyrrole, aniline, and thiophene, leading to the formation of long-chain polyenes. The creation of conjugated polymers begins with a process known as "dimerization," which involves the establishment of a covalent bond between two unsaturated carbon atoms (Liu et al., 2023). The inherent organic characteristics of conjugated polymers ensure their compatibility with biological tissues; if their electronic properties are optimized, this could lead to significant applications in tissue-electronic interfaces and electronic plants. These extraordinary polymers with their unique properties have been noticed as highly promising materials for use in biomedical applications and energy storage (Jadoun et al., 2023; Jadoun et al., 2022; Mozafari and Chauhan, 2019; Pande et al., 2021).

In this chapter, we will outline the synthesis and characterization techniques for conjugated redox polymers. Moreover, we will explore their applications, emphasizing their significance in medicine and energy storage. The highlights are summarized as a diagram in Fig. 1.

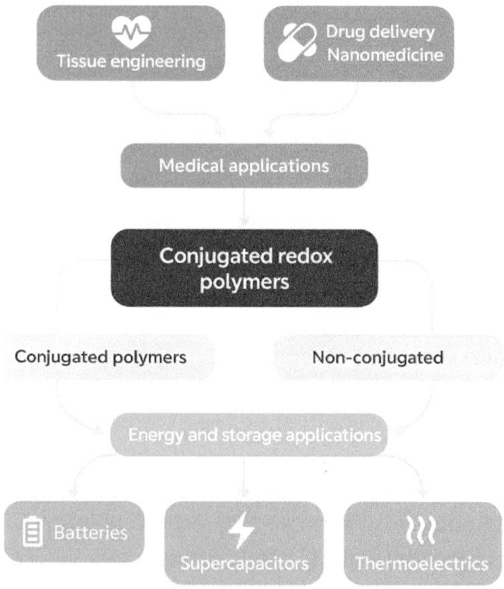

Fig. 1. Applications of Conjugated Redox Polymers.

2. Oxidation-reduction Reactions

Oxidation-reduction (redox) reactions involve the transfer of electrons between species, where oxidation is the loss of electrons and reduction is the gain of electrons. These reactions are characterized by changes in oxidation states and are essential in processes like energy storage, catalysis, and electrochemistry (Casado et al., 2016). A redox reaction might be linked to alterations in the characteristics of the polymers. Consequently, based on their oxidation state (whether oxidized or reduced), polymers typically exhibit varying electronic, chemical, optical, or mechanical characteristics. Redox polymers, such as polyaniline (PANi), polypyrrole (PPy), and poly(3,4-ethylenedioxythiophene) (PEDOT), exemplify redox reactions through their reversible electron transfer capabilities (Fig. 2). For instance, PANi can switch between reduced (leucoemeraldine) and oxidized (pernigraniline) states, altering its conductivity and color. Similarly, PPy and PEDOT undergo redox transitions that enable their use in supercapacitors, sensors, and electrochromic devices. These polymers demonstrate how redox reactions can be harnessed in functional materials for applications such as energy storage, biosensors, electrochromic devices, smart coatings, medicine, and biofuel cells (Deraedt et al., 2014; Kim et al., 2017). Cyclic voltammetry (CV) is a widely used and effective electrochemical technique for investigating the oxidation and reduction reactions of redox polymers. It offers insights into the redox properties of the material across various potentials and serves as a quick and trustworthy characterization method for the confirmation oxidation-reduction process of polymers (Ramya et al., 2013).

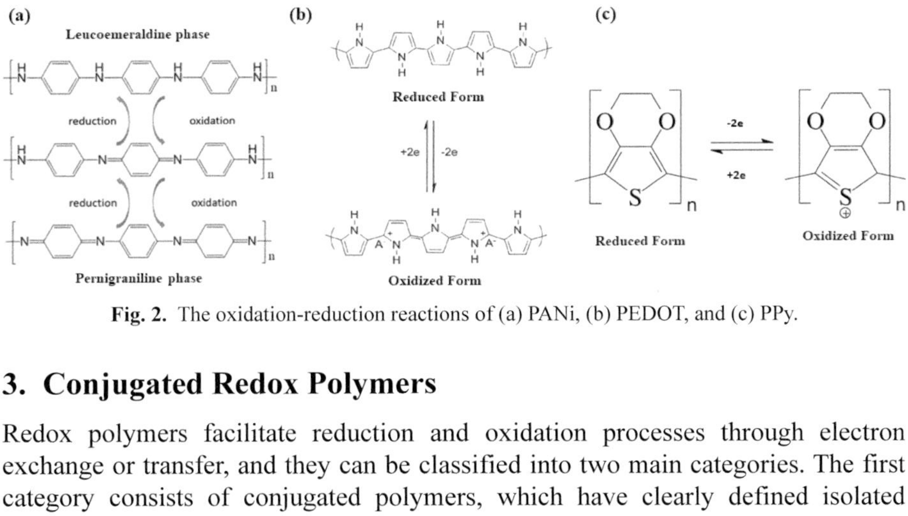

Fig. 2. The oxidation-reduction reactions of (a) PANi, (b) PEDOT, and (c) PPy.

3. Conjugated Redox Polymers

Redox polymers facilitate reduction and oxidation processes through electron exchange or transfer, and they can be classified into two main categories. The first category consists of conjugated polymers, which have clearly defined isolated redox centers. Examples of these include polypyrrole, PANi, polythiophene, and poly (3,4-ethylene dioxythiophene): poly (styrene sulfonate) (PEDOT–PSS) (Table 1) (Chauhan et al., 2021). Conducting polymers (CPs) stand out as exceptional electron-rich materials with the remarkable ability to lose electrons and adopt oxidized states. This innovative process involves the intercalation of anions during oxidation, while reduction results in the withdrawal of these anions, a mechanism that is predominantly associated with cathodes. During discharging, these doped conducting polymers effectively accept electrons, while lithium cations attach seamlessly to the anode (Chauhan et al., 2016; Chundawat et al., 2020a). CPs are known for their capacity to function as highly electron-rich systems, adept at losing electrons and forming oxidized states (Chen et al., 2021; Li et al., 2021). In the charging process, a crucial interaction occurs as anions and lithium cations are efficiently captured by the anode and cathode, respectively, enhancing the functionality and performance of these materials. The second category includes redox polymers that are non-conjugated and contain specific redox moieties, typically attached as side groups. These are the second most significant type of polymeric active compounds (Chundawat et al., 2020b; Yuan and Minteer, 2019). On the other hand, redox polymers are also classified based on the nature of their redox centers, which are either organic or inorganic. This classification underscores the significant differences in properties and applications between the two types (Casado et al., 2016).

We have clarified that Zhou et al. (2023) have significantly advanced the field of Fenton-like catalysis by introducing redox-active polymers, particularly PANi, as electron shuttles to accelerate the Fe^{3+}/Fe^{2+} cycle and enhance H_2O_2 activation. The reduction of Fe^{3+} primarily occurs at the active N–H or O–H bonds through a robust proton-coupled electron transfer process. By serving as effective electron carriers, hydrogen atoms within the solid phase not only inhibit radical quenching but also prevent metal dissolution, thereby maintaining a long-lasting reducing capacity through seamless regeneration. Remarkably, PANi showcases superior redox properties, surpassing other redox-active polymers—including polypyrrole, hydroquinone resin, poly(2,6-diaminopyridine), and the hexaazatrinaphthalene framework—achieving an outstanding iron reduction capacity of 5.5 mmol/g, which represents a transformative shift from leucoemeraldine to emeraldine. Moreover, the innovative system demonstrated exceptional pollutant removal efficiency in a flow-through reactor, effectively operating for 8,000-bed volumes without any noticeable decline in performance. This pioneering work establishes a green and sustainable oxidation system with tremendous potential for advancing practical organic wastewater remediation. The electron transfer processes of redox-active polymers were examined using electrochemical techniques. CV of Fe^{3+} (Fig. 3.a) showed oxidation and reduction peaks at 0.70 and 0.15 V (vs SCE), with an average potential E1/2 of 0.43 V. In contrast, PANi displayed symmetrical peaks at 0.29 and –0.05 V (vs SCE), indicating excellent reversibility. The lower E1/2 of PANi (0.12 V) compared to Fe^{3+} suggested that Fe^{3+} reduction by PANi is thermodynamically favorable. The presence of Fe^{3+} significantly enhanced the current response of PANi,

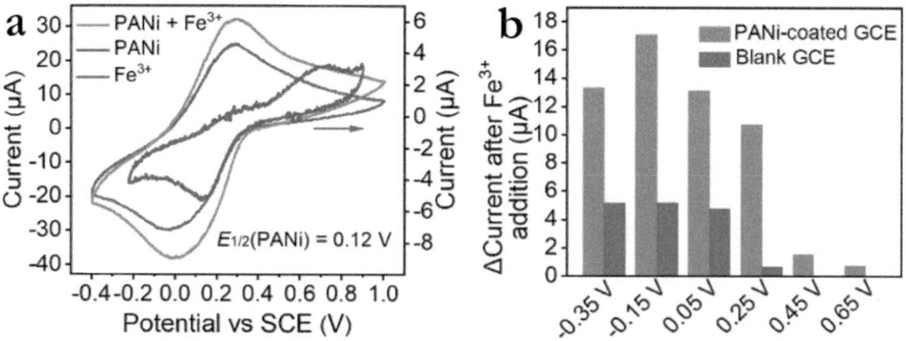

Fig. 3. (a) CV curves of Fe^{3+} alone, PANi alone, and PANi + Fe^{3+} at pH 2.8 (b) Comparison between the current change of the PANi-coated glassy carbon electrode (GCE) and blank GCE at different potentials.

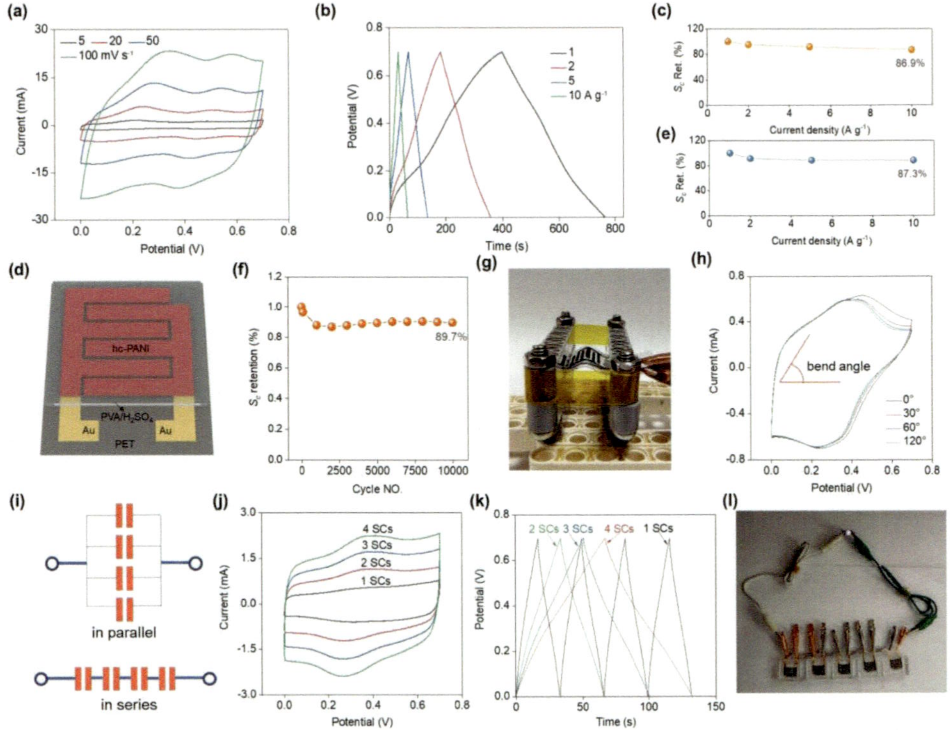

Fig. 4. The remarkable electrochemical properties of hc-PANi. (a–c) hc-PANi supercapacitor electrodes in a three-electrode configuration, (a) CV curve of 5 mV s^{-1}–100 mV s^{-1}, (b) GCD curves at current densities from 1 A g^{-1} to 10 A g^{-1}, (c) Rate-capability of hc-PANi. (d–h) hc-PANi based flexible supercapacitor device, (d) Schematic illustration of supercapacitors, (e) Rate-capability of hc-PANi based flexible supercapacitor device, (f) Capacitance retention after 10,000 cycles, (g) Schematic diagram of flexible test, (h) CV curves of hc-PANi flexible supercapacitors at different bending angles. (i–l) hc-PANi based multiple supercapacitors, (i) Sketch of four supercapacitor devices connected in parallel or in series, (j) CV curves of a single device and multiple devices in parallel, (k) GCD curves of a single device and multiple devices in parallel, (l) The tandem devices light the white LED lamp.

Table 1. The characteristic of conjugated redox polymers.

Polymer	Description	Features	Structure	Ref.
poly(3,4-ethylene dioxythiophene): poly(styrene sulfonate) (PEDOT–PSS)	PSS is an advanced polymer electrolyte that effectively combines PEDOT and PSS. In this mixture, the conductive PEDOT carries positive charges, while the insulating PSS carries negative charges. PEDOT's impressive conductivity is paired with its water insolubility, and PSS allows for its dispersion and stability through Coulombic attractions. Available under reputable brands like Baytron, Bayer AG, and Clevios, these aqueous dispersions of PEDOT-PSS are notable for their striking deep blue opaque hues, making them an excellent choice for various applications.	▪ Tunable conductivity ▪ Processability ▪ Excellent flexibility ▪ Electrochemical performances	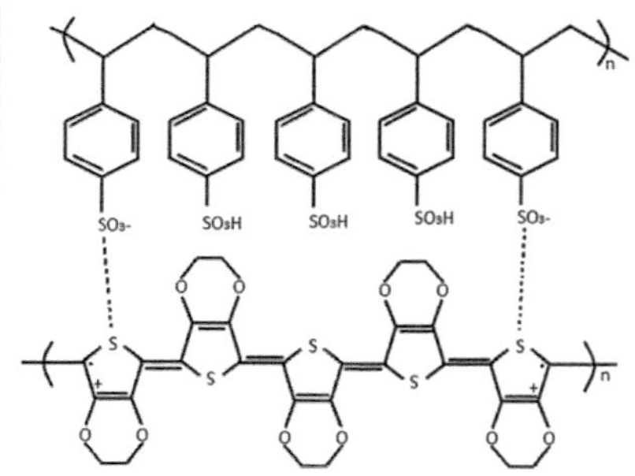	(Vivian R. Feig et al., 2018)
PPy	PPy is a standout supercapacitor electrode material among polymers, ideal for sensors, actuators, batteries, and electrochemical capacitors, thanks to its excellent environmental stability and tunable electrical conductivity. PPy is generally synthesized through the effective method of oxidizing the monomer, either chemically or electrochemically, while also integrating a dopant ion. This dual approach enhances the efficiency of the process and the quality of the final product.	▪ Facile synthesis ▪ Reasonably good redox properties ▪ The intermediate specific capacitance ▪ Better environmental stability ▪ Mechanical flexibility	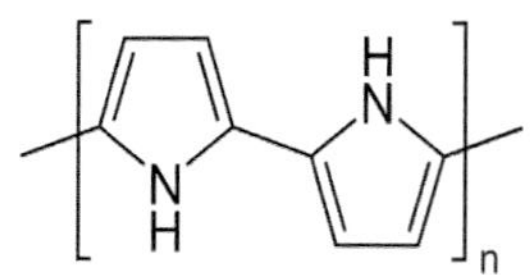	(Huang et al., 2016; Sun et al., 2019)

PANi	Representatives from the family of conducting polymers	▪ Perfect electronic properties ▪ Low cost ▪ Good electronic abilities ▪ Multi-redox reactions ▪ Rationally perfect thermal stability	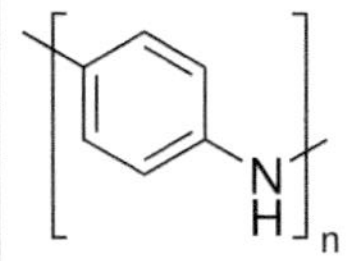	(Boeva and Sergeyev, 2014; Pande et al., 2021)
Polyindole or Polybenzopyrrole (PInd)	Indole-based polymers offer significant potential for electrical energy storage. Various synthesis methods, including electrochemical, microwave-assisted, emulsion, and chemical oxidative polymerization, can be employed to enhance energy storage capabilities.	▪ Strong redox ability ▪ Variable electrical properties ▪ Significant heat resistance ▪ Moderate degradability	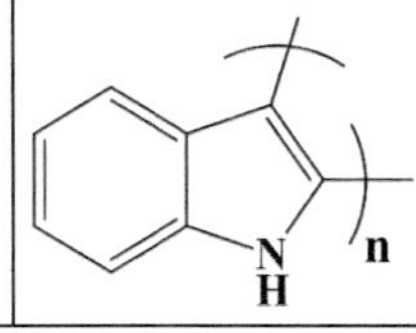	(Marriam et al., 2020)

indicating electron transfer. Chronoamperometry analysis (Fig. 3.b) revealed that PANi-coated glassy carbon electrodes (GCE) showed higher current responses after Fe^{3+} addition, although the current change diminished with increasing potential.

Wang et al. (Yihan Wang et al., 2023) present groundbreaking findings on the hyper-conjugated effect, which significantly lowers energy level differences and effectively reduces the barrier for carrier migration in copolymerized PANi, a leading supercapacitive material. This powerful effect not only elongates the conjugated system but also enhances the delocalization of π electrons, resulting in a remarkable increase in the number of free carriers. The expanded range of carrier mobility leads to an impressive 800% surge in electrical conductivity. Due to both the hydrogen-bonding effect and the hyper-conjugated effect, hc-PANi materials exhibit a remarkable specific capacitance of 521 F g^{-1} at a current density of 1 A g^{-1}, which is 50% higher than that of p-PANi. Furthermore, flexible supercapacitors based on hc-PANi demonstrate excellent stability, retaining 87.3% of their specific capacitance even when the current density increases to 10 A g^{-1}. After 10,000 cycles, these supercapacitors maintain a capacitance retention of 89.7%, making them suitable for wearable electronics and various other applications (Fig. 4).

In Table 1, additional polymers along with their characteristics and structures are listed.

4. Synthesis Techniques and Characteristics of Conjugated Redox Polymers

The synthesis methods used to prepare conjugated polymers significantly influence their physical and chemical properties, such as conductivity, morphology, and electrochemical stability. These key properties, in turn, directly affect the overall performance of supercapacitors (SCs) that employ polymers as electrode materials. Therefore, selecting an appropriate synthesis method is critical for achieving the desired characteristics required for high-performance SCs. Polymers used in SCs are synthesized through various methods, including electrochemical polymerization (electropolymerization), chemical polymerization, and in situ polymerization. This section discusses the synthesis of polymers via these three approaches and evaluates their electrochemical performance (Genene et al., 2024). Additionally, in the last subtitle, the techniques used to characterize redox-conjugated polymers will be summarized.

4.1 Electrochemical Polymerization

Electropolymerization (EP), a prevalent electrochemical synthesis technique, is widely employed for fabricating conjugated polymers due to its capacity to concurrently synthesize and deposit polymer films onto conductive substrates in a single-step process. Electrochemical polymerization provides certain benefits compared to chemical polymerization, including the lack of a catalyst and the ability to regulate the film's thickness and morphology. The mechanism proceeds through three distinct stages: (i) electrochemical oxidation of monomers to generate cation-radicals, (ii) radical coupling of these intermediates, and (iii) rearomatization via deprotonation (Heinze et al., 2010). This sequence culminates in oligomer formation, followed by film deposition driven by nucleation and polymer chain elongation. EP facilitates direct polymer adhesion to charge collectors, circumventing the need for binder additives that often diminish conductivity while producing electrodes with high porosity and extended surface areas. These structural attributes enhance ion diffusion kinetics and pseudocapacitive charge storage in SCs. Nevertheless, the method faces challenges, including incompatibility with monomers possessing high oxidation potentials, complexities in determining polymer molecular weights, and self-limiting growth mechanisms that restrict achievable mass loading. Homopolymers, such as PANi and PPy, represent conventional conjugated polymers extensively utilized in electrochemical energy storage systems. Their prominence stems from high electrical conductivity, substantial pseudocapacitance derived from reversible redox activity, and cost-effective synthesis routes (Genene et al., 2024). Recent advancements in EP have further optimized these materials, yielding electrodes with improved cycling stability and mechanical flexibility. For instance, Yigit and Güllü synthesized carbazole-based homopolymers (P7 and P8) via electropolymerization of monomers functionalized with thiophene and ethylenedioxythiophene (EDOT), respectively. These polymers exhibited enhanced cycling stability in SCs, attributed to their robust redox activity. During charge/discharge cycles, polarons form within the polymer matrix, enabling energy storage through Faradaic redox reactions localized at the electrode-electrolyte interface, as depicted in Fig. 5 (Yiğit and Güllü, 2017).

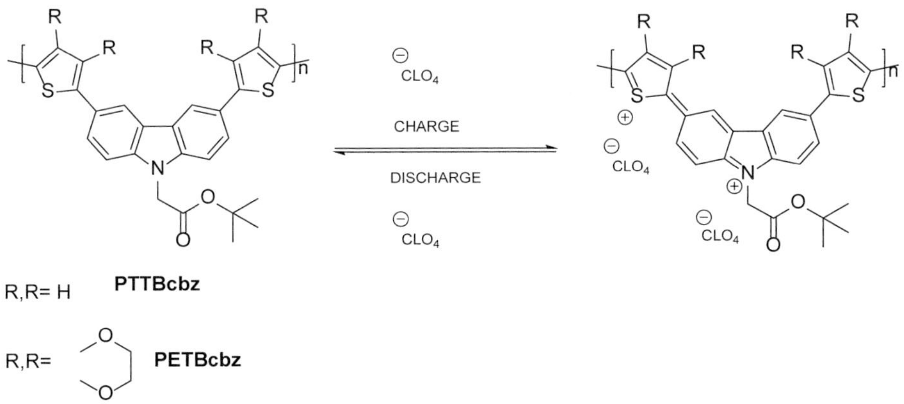

Fig. 5. Faradaic redox mechanisms of P7 and P8 (Charge/Discharge mechanism of PTTBCbz and PETBCbz redox-active materials).

4.2 Chemical Polymerization

In contrast to EP, chemical polymerization (CP) employs oxidative agents or transition metal-catalyzed cross-coupling reactions to synthesize donor-acceptor (D-A) polymers with tunable electronic structures. This method is particularly advantageous for the scalable production of solution-processable p-type and n-type polymers, rendering it highly suitable for industrial applications. However, these benefits are accompanied by notable limitations, including reliance on toxic reagents, the complexity of multistep synthesis protocols, and lower energy density compared to EP-derived materials (Malik et al., 2023). To illustrate these trade-offs, Wu et al. demonstrated the method's potential by fabricating a nanostructured PANi/reduced graphene oxide (PANi/rGO) composite gel. This composite was synthesized via a solution-assisted two-step self-assembly method, starting from graphene oxide (GO) and chemically polymerized PANi. The resulting PANi/rGO electrode exhibited a high specific capacitance (Cs) of 824 F g^{-1} (5830 mF cm^{-2}) at 2.22 A g^{-1} (15.72 mA cm^{-2}) and displayed excellent rate performance. These outcomes were attributed to the composite's favorable morphology, which enhanced electrolyte diffusion and electron transport pathways (Wu et al., 2018).

Expanding on chemical polymerization's versatility, transition metal-catalyzed cross-coupling reactions such as Suzuki polymerization, Stille polymerization, and direct (hetero)arylation polymerization (DHAP) enable precise coupling of monomers in the presence of metal catalysts. Notably, these techniques yield solution-processable p-type and n-type polymers that can be deposited onto flexible, transparent substrates like indium tin oxide (ITO)-coated polyethylene terephthalate (PET) (Jean-Rémi et al., 2016). For industrial applications, such polymers are compatible with scalable deposition methods, including spray coating, spin coating, and roll-to-roll printing. Despite challenges like toxicity and process complexity, this adaptability underscores chemical polymerization's dual role in advancing energy storage materials and flexible electronics, positioning it as a critical strategy for balancing performance, scalability, and functional design.

4.3 In Situ Polymerization

The integration of conjugated polymers with functional nanomaterials provides a strategic approach to overcoming the limitations inherent in individual material systems. By optimizing synthesis protocols such as combining the binder-free deposition capabilities of electrochemical polymerization (EP) with the composite compatibility of in situ polymerization researchers can design electrodes with hierarchical porosity, efficient charge transport pathways, and improved electrochemical stability (Vijayakumar et al., 2021; Xie et al., 2016). A notable example of this strategy is the synthesis of stable sulfonated lignin and PEDOT biocomposites (LG/PEDOT) via *in situ* chemical polymerization and electrochemical polymerization, as shown in Fig. 6. This work emphasized the critical role of incorporating a hydroquinone/quinone redox couple into the lignin structure, which significantly enhanced charge storage performance. In this system, lignin functioned as both a dopant and a surfactant, facilitating the development of sustainable organic electrode materials for energy storage applications. Specifically, PEDOT doped with lignin achieved a specific capacitance of 170.4 F g^{-1}, twice that of reference PEDOT electrodes (80.4 F g^{-1}). This enhancement was attributed to the additional pseudocapacitance provided by

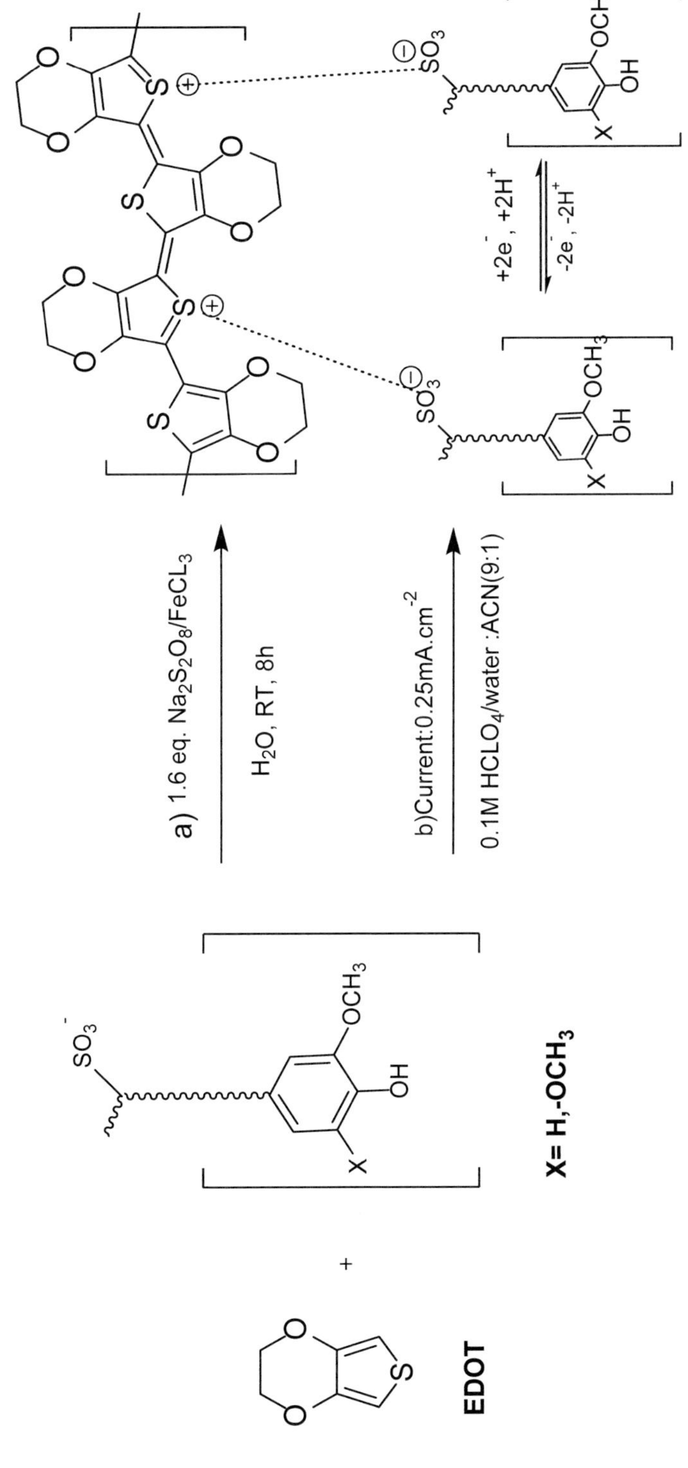

Fig. 6. (a) Chemical oxidative polymerization and (b) electrochemical polymerization of the PEDOT/Lig composite.

the quinone moieties in lignin, which enabled faradaic redox reactions. Furthermore, the PEDOT/lignin biocomposite demonstrated exceptional electrochemical stability, retaining approximately 83% of its electroactivity after 1000 charge/discharge cycles. These results highlight the effectiveness of redox doping as a simple yet robust strategy for improving the electroactive performance of PEDOT-based materials, presenting significant potential for advancing energy storage technologies (Ajjan et al., 2016).

4.4 Characterization Methods

In the synthesis of polymers, the kinetics, especially the graph of molecular weight to conversion, significantly vary between polycondensation (step-growth) and polymerization (chain-growth) methods. A significant outcome of conjugated polymer synthesis through polycondensation is the need to reach high conversion, as this is essential for obtaining high molecular weights. The impact of the loss of functional groups on preserving proper stoichiometries has been noted, and this can consequently restrict the molecular weights of those polymerizations. One more outcome pertains to the leftover end groups. The functional units (such as halo, boronate, or others) can significantly interfere with device performance and must be eliminated through specific end-capping. Nevertheless, this situation becomes even more challenging if it is difficult to precisely identify the end group (Qiu et al., 2020). So, it is necessary to characterize the synthesized polymers. A major challenge associated with redox polymers is accurately characterizing them to enhance the understanding of the relationship between structure and properties. In addition to traditional polymer characterization methods for assessing molecular weight distribution, physicochemical attributes, glass transition, crystallization, thermal, and mechanical properties, the redox characteristics are typically analyzed through electrochemical techniques. Besides the inherent characteristics related to the type of polymer, the external factors involved in sample preparation and the electrochemical methods used are also essential for characterizing redox polymers. A good example of characterization methods for a conjugated redox polymer is discussed earlier in Title 3, Fig. 4(a–c). The primary external factors that must be considered for the characterization of redox polymers include: (i) the condition of the sample, (ii) the kinetics of the redox reaction, (iii) the type of electrolyte used, (iv) the temperature at which the experiment is conducted, (v) the kind of electrochemical cell employed, and (vi) the composition of the electrode when incorporating binders and conductive materials (Casado et al., 2016). A summary of all the characterization methods can be found in Table 2.

Table 2. Techniques for verifying the structure of redox conjugated polymers.

Characterization Technique	**Purpose**	**Definition/Key Findings**
UV-Vis-NIR Spectroscopy	Electronic structure analysis	It is represented as the photon intensity that persists following the irradiation of the sample.
FTIR	Functional group identification	Confirmed the interaction between functional groups of polymer, additive, or agents.
Thermogravimetric Analysis (TGA)	Thermal stability assessment	Demonstrated enhanced thermal stability and measured the weight loss.
SEM/TEM	Morphological analysis	Revealed homogeneous composite structure with porous architecture.
CV/GCD	Electrochemical performance evaluation	Quantified doubled specific capacitance and redox activity.
Electrochemical Impedance Spectroscopy (EIS)	Electrical property evaluation	The electrical properties of the polymer films were examined using a robust and intricate method.
Cycle Stability Testing	Long-term durability assessment	Measured capacitance retention after N cycles.

5. Medical Applications

Conjugated redox polymers form an emerging field that uses electricity to regulate and interact with biological tissues, especially electrogenic tissue, delivering high-temporospatial-resolution and personalized therapy. They are increasingly studied for various medical applications due to their unique properties, including electrical conductivity, biocompatibility, and the ability to undergo redox reactions. These features make them ideal candidates for advancing healthcare solutions. Here are the key areas of application:

5.1 Tissue Engineering

Redox-conjugated polymers are engineered to form optimal structures that facilitate cell attachment and growth. Their unique electrical properties not only boost cell proliferation but also enhance differentiation, making them essential for the successful engineering of critical tissues, such as cardiac and neural tissues. In recent years, tissue engineering has made significant advancements, particularly in the development of new materials and fabrication techniques for scaffolds and conduits needed to create functional tissues. Key properties for an optimal scaffold system include tailoring physicochemical and mechanical properties, effective delivery of bioactive agents, and constructing complex architectures with spatially arranged cues. While these methods offer important stimuli for guiding biological interactions, they often overlook the crucial role of electrical stimuli in cellular behavior. Research has shown that electrical stimulation fosters essential biological processes such as cell adhesion, proliferation, and differentiation, revealing vast potential in tissue engineering (Dubey et al., 2011; Kotwal and Schmidt, 2001; Tandon et al., 2009). However, traditional metal electrodes are often unsuitable due to issues like rigidity and non-degradability. The discovery of organic conducting polymers (OCPs) in the late 1970s by Shirakawa, MacDiarmid, and Heeger (Chiang et al., 1977; Shirakawa et al., 1977) opened new avenues for research, leading to applications in flexible electronics and eventually in biological fields. OCPs are valued for their biocompatibility and customizable properties, leading to their use in biosensing (Inal et al., 2018; Xia et al., 2010), drug delivery and imaging (Sarkar and Levi-Polyachenko, 2020), tissue engineering (Biglari and Zare, 2024; Sharifabad et al., 2025), and bio-electronics interface (Parashar et al., 2020). The ability of OCPs to undergo redox processes and release incorporated ions during electrochemical polymerization makes them ideal for controlled drug delivery systems. Additionally, these materials show great potential in neural tissue engineering, advancing medical technology (Thompson et al., 2011). OCPs such as PPy, PANi, and PEDOT represent a groundbreaking class of materials that excel in biocompatibility and hold immense potential for various biomedical applications. These innovative materials have demonstrated their ability to promote guided cell growth (Breukers et al., 2010), facilitate controlled drug delivery (Richardson et al., 2007), provide targeted electrical stimulation (Thompson et al., 2010), enable effective recording (Harris et al., 2012), and offer mechanical stimulation (Gelmi et al., 2016) all critical functions in advancing healthcare solutions. OCPs are characterized as conjugated polymers, where the oxidation of the monomeric unit—via electrochemical or chemical methods—results in a positively charged polymer backbone. This charge is balanced by incorporating an anionic species, known as a dopant, in a vital process called doping. By harnessing the unique properties of OCPs, we can drive significant advancements in medical technology and improve patient outcomes (Wallace et al., 2008).

These polymers offer innovative and versatile methods for fabricating 2D films, employing techniques such as vapor phase deposition (Stewart et al., 2013), electrochemical polymerization (Molino et al., 2018), and spin coating (Zhang et al., 2015). Moreover, their application extends to inks and composites, enabling advanced manufacturing through electrospinning (KarbalaeiMahdi et al., 2022; Prabhakaran et al., 2011), wet spinning (Foroughi et al., 2011), and 3D printing (Sharifabad et al., 2025). The other applications and details are summarized in Table 3.

Scaffolds via conjugated polymers can be fabricated using methods such as solution casting, electrospinning, *in-situ* polymerization, and melt blending. The choice of method depends on the types of polymers and fillers, synthetic conditions, and intended applications, each offering distinct advantages and challenges (Kausar, 2021). Melt blending is an effective and widely used method in polymer synthesis for incorporating conductive elements into a viscous polymer melt (see Fig. 7.a). It offers key advantages, such as preventing filler re-aggregation, eliminating the need for chemical modifications, and allowing for immediate filler dispersion. When the material concentration exceeds 5%, costs can increase due to higher melt viscosity (Fig. 7.b), making solution casting a crucial alternative. This technique utilizes a solvent to dilute nanofillers, ensuring excellent dispersion within the

Table 3. Tissue engineering applications of conjugated redox polymers.

Tissue type	Structure	Material	Insights and Potential	Ref.
Cardiac tissue	Film on chitosan	PANi -phytic acid	*Ex vivo* experiments demonstrate that the patch's conductive properties immediately affect the heart's electrophysiology. Moreover, preliminary in vivo studies confirm it does not induce proarrhythmogenic activities.	(Mawad et al., 2016)
	Electrospun mat	PPy/PCL/Gelatin	PPG15 nanofibers, which include 15% PPy, excel in mechanical strength, electrical conductivity, and degradation properties for cardiac regeneration, outperforming PG and PPG30 scaffolds. Cardiomyocytes on PPG15 scaffolds show enhanced adhesion, proliferation, and cardiac-specific protein expression.	(Kai et al., 2011)
		PANi -PLGA	Every cardiomyocyte in each cluster beats in harmony, showing the cells are well connected. Additionally, we can synchronize their beating rates using electrical stimulation that mimics natural heart signals, highlighting their potential for therapeutic applications in heart health.	(Hsiao et al., 2013)
		PANi -PLA	Conductive nanofibrous sheets provide an exceptional biocompatible microenvironment that enhances the viability, maturation, and synchronized beating of cardiac myocytes (CMs). The PLA/PANi conductive nanofibrous sheets emerge as promising 3D bioactuators for CMs. Their ECM-like structure and inherent conductivity not only improve biocompatibility but also significantly enhance H9c2 cell differentiation, resulting in increased myotube formation and maturation.	(Wang et al., 2017)
	Hydrogel	PPy-chitosan	PPy-chitosan is a groundbreaking biocompatible conductive biomaterial that enhances biological conduction both in vitro and *in vivo*. It effectively maintains heart function after myocardial infarction, outperforming non-conductive polymers.	(Mihic et al., 2015)
		Chitosan-*graft*-aniline tetramer and dibenzaldehyde-terminated PEG	The C2C12 myoblasts and H9c2 cardiac cells exhibit an adaptable release rate, with strong *in vivo* cell retention in conductive hydrogels. Studies on subcutaneous injection and degradation confirm their injectability and biodegradability. These self-healing, conductive, and biodegradable injectable hydrogels are promising candidates for advanced cell delivery vehicles in cardiac repair.	(Dong et al., 2016)

Table 3 contd. ...

...Table 3 contd.

Tissue type	Structure	Material	Insights and Potential	Ref.
Skeletal muscle tissue	Co-polymer film	Amino capped aniline trimer/PCL	These shape memory copolymers demonstrate remarkable strain at break, impressive electroactivity, and effective shape memory properties with a recovery temperature around body temperature. They exhibit high recovery and fixity ratios, along with quick recovery times. Notably, our cell studies show that these electroactive PCL3000-AT copolymers significantly enhance C2C12 myoblast proliferation compared to PCL80000. They also promote myogenic differentiation, leading to improved myotube formation and gene expression levels. This research suggests that these flexible copolymers, with recovery temperatures near or slightly above 37°C, are excellent candidates for scaffolding biomaterials in muscle tissue engineering.	(Deng et al., 2016)
	Composite film	PPy/polyurethane	PPy nanoparticles and polyurethane (PU) composites demonstrate excellent elastomeric properties, conductivity, and compatibility with C2C12 cells. Their electrically conductive network within the elastomer positions these materials as promising electromechanical couplers for advancing cellular applications in tissue engineering.	(Broda et al., 2011)
		Aniline trimer/ polyurethane-urea co-polymer	The copolymers show excellent solubility in various organic solvents, indicating superior processability compared to PANi. They also exhibit greater thermal stability than pristine PLLA. PUU copolymer films are non-cytotoxic and effectively support the adhesion and proliferation of C2C12 myoblast cells. Enhanced myogenic differentiation is evident through morphological observations and quantitative analysis of myotubes and gene expression. These results underscore the significant potential of electroactive, elastic, and degradable PUU copolymers in soft tissue engineering applications where electroactivity is essential.	(Chen et al., 2015)

	Electrospun mat	PANi/PLCL	PLCL/PANi fibers may have a limited effect on myoblast proliferation, but they significantly enhance myogenic differentiation. Cells in these fibers showed increased numbers and lengths of myotubes, along with elevated expression of key myogenic genes like myogenin, troponin T, and MHC, highlighting the potential of PANi fibers in promoting muscle development.	(Jun et al., 2009)
		PANi/PCL	C2C12 myoblasts cultured on aligned nanofibers (A-PCL and A-PCL/PANi) exhibited oriented cellular morphologies, which significantly enhanced differentiation compared to R-PCL scaffolds. The addition of PANi to the PCL scaffolds proved non-toxic to the myoblasts, ensuring safe growth. Furthermore, the electrical properties of A-PCL/PANi nanofibers stimulated the formation of multi-nucleated myotubes, demonstrating their superior benefits over non-conductive A-PCL scaffolds.	(Chen et al., 2013)
Bone tissue engineering	Composite film	Aniline trimer/PLA	The SMP (PH10 and PH20) and ESMP showcase remarkable shape memory properties, including rapid recovery time, high recovery ratio, and exceptional fixity ratio. Notably, the cell proliferation assay revealed that the electroactive ESMP not only demonstrates excellent biocompatibility but also significantly boosts the proliferation of C2C12 cells compared to traditional SMPs and PLA85k. Furthermore, the ESMP substantially enhances the osteogenic differentiation of C2C12 cells when stacked against PH10 and PLA85k.	(Xie et al., 2015)
	Composite membrane	PPy/He/PLLA	An optimal direct current (DC) potential gradient of 200 mV/mm for 2 to 8 hours significantly enhances osteoblast proliferation and the expression of key markers for the maturation (ALP) and mineralization (OC) stages of bone formation. These findings highlight the innovative potential of combining electrical stimulation (ES) with conductive scaffolds in bone tissue engineering.	(Meng et al., 2013)

Table 3 contd. ...

...Table 3 contd.

Tissue type	Structure	Material	Insights and Potential	Ref.
	Electrospun scaffold	PANi/PLA	Live/dead cell staining and scanning electron microscopy (SEM) observations demonstrate that the composite scaffolds support excellent cell adhesion and are non-cytotoxic. Compared to pure polylactic acid (PLA) scaffolds, the conductive composite scaffolds significantly promote cell proliferation, osteogenic differentiation, and calcium mineralization, especially in the PLA/PANi 10 sample with optimal PANi content. With nanofibrous three-dimensional structures and bone-like conductivities, these PLA/PANi composite scaffolds are promising biomaterials for bone tissue regeneration.	(Chen et al., 2018)
	Porous scaffold	PPy/SiO_2/gelatine/ hydroxyapatite	PPy is an ideal material for bone tissue engineering, showing no toxicity to osteoblast cells. It combines essential mechanical strength with a conductive structure that enables drug release over four months from a bone scaffold, enhancing both bone regeneration and treatment effectiveness.	(Zanjanizadeh Ezazi et al., 2018)
	Scaffold	PEDOT/PCL	PCL scaffolds are highly effective as *in vivo* bone regeneration grafts, while PEDOT stands out in organic bioelectronics for its stability, flexibility, and biocompatibility. These scaffolds maintain essential porosity and mechanical stability for bone regeneration. Additionally, the proliferation of human fetal mesenchymal stem cells on the functionalized scaffolds demonstrates the excellent cytocompatibility of the polymeric coating, highlighting their potential in regenerative medicine.	(Iandolo et al., 2016)
Skin tissue	Chemical polymerisation on fabric	PPy modified PET fabric	PES significantly boosted TGFβ1 production in human dermal fibroblasts, playing a crucial role in their transformation. This study shows that TGFβ1 regulates the transdifferentiation of these cells into myofibroblasts through the TGFβ1-ERK-NF-κB signaling pathway. Importantly, the PES-induced phenotype change persists after five days of in vitro cell expansion and 15 days post-grafting, highlighting its potential for long-term therapeutic use.	(Yongliang Wang et al., 2016)
	Composite film	PPy/PLA	The newly developed conductive PPy/PLLA membranes significantly enhance the attachment, spreading, and proliferation of human cutaneous fibroblasts. A constant DC electrical field results in a remarkable 2.2 to 4.0 times increase in cell viability compared to non-stimulated cultures over 2 to 24 hours. Interestingly, a strong DC current does not affect viability due to its minimal potential gradient.	(Shi et al., 2008)

		PPy/Heparin/PLLA	PPy/PLLA membranes with heparin as a dopant demonstrate enhanced electrical stability in aqueous environments and improved adhesion and growth of human fibroblasts. These properties position PPy/PLLA composites as promising candidates for biomedical engineering applications.	(Meng et al., 2008)
	Electrospun mat	PANi/CPSA/PLCL	The growth of NIH-3T3 cells on CPSA-PANi/PLCL scaffolds is significantly enhanced through moderate electrical stimulation. These innovative nanofiber scaffolds offer exceptional mechanical strength, electrical conductivity, and nanometer-scale features resembling a native extracellular matrix (ECM).	
		PANi -chitosan	The nanofibrous membranes incorporating varying ratios of PANi and chitosan demonstrate increased contact angles correlated with PANi content, resulting in a more hydrophobic surface. Impressively, none of the membranes showed cytotoxicity; instead, they supported strong cell attachment and proliferation, even at high PANi levels, contradicting previous literature findings. These positive results suggest that further investigation into fibroblast and osteoblast responses to electrical stimuli could lead to significant advancements in biomaterial applications.	(Moutsatsou et al., 2017)
Neural tissue	Electrochemical polymerised film	PEDOT doped with Ulvan, DS, CS, ALG or DBSA	Ulvan, an innovative algal-derived dopant, is a promising choice for enhancing PEDOT films due to its excellent biocompatibility and electrochemical properties. Electrical stimulation of PC-12 cells on PEDOT-DBSA and PEDOT-ULV significantly boosts cell differentiation, highlighting the potential of using large, naturally occurring polysaccharides as dopants in PEDOT biomaterials for advanced electrical stimulation of cells and tissues.	(Molino et al., 2018)
		PPy-CS-Collagen	They have developed a novel method that utilizes functional groups on a large polyelectrolyte ECM dopant to anchor secondary biomolecules. This allows for the in situ formation and covalent bonding of a collagen fibrillar matrix on a conducting polymer's surface. This technique creates a robust three-dimensional ECM structure, surpassing simple molecular coatings at the cellular-electrode interface. Additionally, it preserves the polymer's electrical properties by targeting only the excess dopant at the surface, promoting effective self-assembly of the collagen matrix.	(Liu et al., 2011)

Table 3 contd. ...

...Table 3 contd.

Tissue type	Structure	Material	Insights and Potential	Ref.
		PPy-PSS-NGF	Electrical stimulation with PPy-NGF enhanced neurite length compared to unstimulated controls. This underscores the potential synergy of electrical and chemical stimuli from the material's bulk and surface properties. Additionally, improved cell adhesion resulted from the positively charged PAA and increased roughness from the immobilization process, creating a better environment for cell growth.	(Gomez and Schmidt, 2007)
	Electrospun mat	PANi/PCL/Gelatin	The concentration of PANi in the biocomposite significantly enhances the conductivity and biological properties of electrospun nanofibrous scaffolds. Electrical stimulation of neural stem cells (NSCs) through these scaffolds boosts cell proliferation and neurite outgrowth compared to non-stimulated ones.	(Ghasemi-Mobarakeh et al., 2009)
		PANi/PLL	Electrical stimulation of nerve stem cells on conducting polymer scaffolds of PLLA and PANi promotes differentiation and neurite elongation, highlighting the crucial role of electrical impulses in nerve tissue regeneration.	(Prabhakaran et al., 2011)
	Electrospun nerve conduit	PPy-pts modified PLCL/SF	The SFI values studies showed that the Ppy-coated NGC group had better recovery of sciatic nerve function, highlighting the Ppy-coated PLCL/SF NGC as an effective option for enhancing peripheral nerve tissue regeneration.	(Sun et al., 2019)
	Hydrogel	PEDOT/chitosan/gelatin	The assessments revealed that the PEDOT-assembled Cs/Gel scaffold significantly boosts PC12 neurite growth, leading to higher protein and gene expression levels.	(Wang et al., 2017)
		PEDOT-CMC	The integration of a conductive PEDOT network greatly enhances the electrical and mechanical properties of hydrogels while reducing porosity, water absorption, and biodegradability. After nine days of culture, hydrogels with a PEDOT layer effectively supported the spreading and proliferation of PC12 cells, even without electrical stimulation. Additionally, the roughened surface of the PEDOT layer improves cell adhesion, highlighting the potential of these hydrogels for advanced biomedical applications.	(Xu et al., 2018)

	Aerogel	PPy-cellulose aerogel	In vitro studies revealed that adding PPy to a nanocomposite gel (NCG) enhances the adhesion and growth of PC12 cells. Notably, electrical stimulation showed that these cells not only adhered better but also developed longer neurites on NCG/PPy composite gels with a dodecylbenzenesulfonic acid (DBSA) dopant, showcasing its promise for neural tissue engineering.	(Shi et al., 2014)
	Conduit	PEDOT/agarose	This innovative lining promotes axonal growth and serves as a vital interface between neural prosthetic devices and the peripheral nervous system, improving functionality and patient outcomes.	(Abidian et al., 2012)

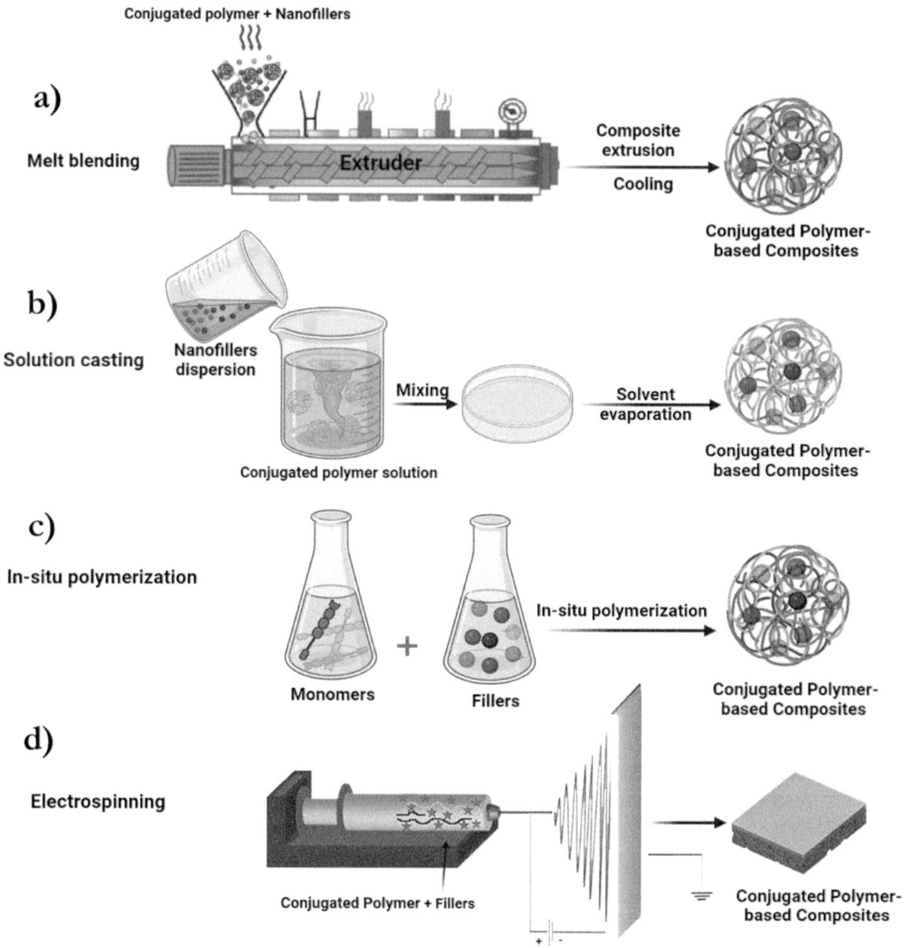

Fig. 7. Innovative fabrication techniques for developing advanced nanocomposites based on conjugated polymers.

polymer. Solution casting is a powerful method for dispersing conductive fillers effectively (Fig. 7.c). It involves dissolving fillers in a monomer solution and adding a polymerization initiator to form robust composites. Additionally, electrospinning offers an accessible and cost-effective way to produce continuous, randomly oriented, or uniaxially aligned fibers from polymer melts or solutions (Fig. 7.d) (Kausar, 2021).

5.2 Drug Delivery and Nanomedicine

Conjugated redox polymers offer a groundbreaking approach to drug delivery by releasing therapeutic agents in a precisely controlled manner. With the ability to respond to physiological stimuli, these innovative materials facilitate targeted treatment while significantly reducing side effects, making them an ideal solution for effective medication management (Molino et al., 2021). The importance of this system arises from the significant difference in redox potential between the oxidizing extracellular environment and the reducing intracellular environment. The cytosol has a low redox potential due to high levels of reduced glutathione (GSH), typically around 10 mm inside the cell, compared to only 0.002 mm outside (Urban, 2011). In tumor cells, GSH concentrations exceed those found in healthy tissues. This significant difference in redox potential presents a promising opportunity for innovative redox-responsive intracellular delivery systems. Several specially designed redox-responsive polymers have been effectively utilized, paving the way for precise and controlled drug release strategies that can enhance treatment outcomes (Huo et al., 2014).

A redox-responsive polymer based on poly(ethylene glycol)-b-poly(lactic acid) (MPEG-SS-PLA) diblock copolymers have been developed by Song et al. (Song et al., 2011) These copolymers self-assemble into rice-grain-shaped polymeric nanoparticles (PNPs). *In vitro* studies show that paclitaxel-loaded PNPs release the drug rapidly under reducing conditions, outperforming redox-insensitive MPEG-PLA PNPs. Additionally, these PNPs exhibit low cytotoxicity, and high cytocompatibility, and enhance the endocytosis of antitumor drugs against MCF-7, HeLa, and A549 carcinoma cells.

In another approach, Zhao et al. introduced innovative redox-responsive nanocapsules for protein delivery by encapsulating Caspase-3 using *in situ* interfacial polymerization within a positively charged polymeric shell connected by disulfide crosslinkers (Zhao et al., 2011). When exposed to GSH, these systems effectively released the protein, inducing apoptosis in human cancer cell lines like HeLa, MCF-7, and U-87 MG. Additionally, RAFT polymerization has been crucial in developing redox-responsive thiol-containing drug delivery systems, showcasing its potential in targeted cancer therapies. Pyridyl disulfide-functionalized RAFT agents are revolutionary in mediating the polymerization of a wide range of molecules, resulting in the innovative development of pyridyl disulfide-functionalized amphiphilic block copolymers. These versatile end groups can be efficiently conjugated to thiol-containing biomolecules, such as peptides, carbohydrates, and proteins, opening new avenues for research and applications in biotechnology and materials science (Jia et al., 2009). This innovative approach enables the development of redox-responsive biomolecule–polymer conjugates. In a groundbreaking study by Boyer et al., cationic star polymers were skillfully engineered as advanced delivery vehicles for siRNA, effectively targeting and silencing disease-causing genes (Boyer et al., 2013). By utilizing dimethylaminoethyl methacrylate (DMAEMA) as a monomer in RAFT polymerization, and incorporating both N,N-bis(acryloyl) cystamine and DMAEMA as crosslinkers, we have successfully created well-defined biodegradable star polymers. These innovative polymers effectively self-assemble with siRNA, resulting in the formation of spherical nanoparticles measuring 35 nm in diameter. This size is particularly advantageous for drug delivery applications, highlighting the potential of these materials in advancing therapeutic efficacy.

6. Energy and Storage Applications

Throughout history, humanity has sought to improve living standards for future generations. Today, in pursuing a better life and advanced technologies, we face a critical energy crisis and growing environmental pollution. The connection between these issues is undeniable, particularly as electrical energy becomes increasingly central to our lives. With rising demand for innovative electronic devices and electric vehicles, we must focus on developing efficient energy conversion, storage, and utilization technologies. Addressing these challenges is essential for ensuring a sustainable future for all.

Energy storage systems (ESSs) are vital for storing energy from traditional power sources, as well as renewable sources like solar, tidal, wind, and waste energy. They power electric vehicles and personal electronics such as laptops and smartphones and are essential for developing future portable devices like smart tags and wearable technology (Lewis, 2016). To meet future demands, ESSs must offer high performance, superior storage capacity, fast charging, enhanced safety, long lifespan, versatility, lightweight design, and cost-effectiveness while being non-toxic. These challenges are ongoing tasks for materials scientists and engineers (Larcher and Tarascon, 2015). Current ESSs like supercapacitors and rechargeable batteries predominantly rely on metal-based inorganic compounds (e.g., Fe, Co, Ni). However, these materials raise toxicity and environmental concerns, and their performance is limited by the specific counter-ions tied to their crystal structures (Park et al., 2015). Research into next-generation ESSs, including advanced supercapacitors and various types of rechargeable batteries, has faced challenges concerning stability and safety. Therefore, innovative approaches must be pursued to address the limitations of existing ESSs and enhance their performance for future applications (Stamenkovic et al., 2017).

Research on energy storage devices that utilize sustainable bio-inspired materials has been very active in recent years (Zhang et al., 2015). To enhance energy storage solutions, we must prioritize redox-active materials for next-generation ESSs. These materials offer significant advantages, including optimal molecular structures, superior properties, and exceptional electrochemical performance compared to traditional metal-based materials (Schon et al., 2016). Utilizing redox-active materials can greatly advance current ESS technologies and foster innovative solutions. Redox-active polymers, particularly, are favored for ESS applications due to their stability, processability, and ease of fabrication (Casado et al., 2016). Conjugated redox polymers exhibit remarkable versatility and electrochemical properties, positioning them for a wide range of applications. Here are the key applications:

6.1 Batteries

Conjugated redox polymers serve as exceptional cathode materials in both lithium-ion and sodium-ion batteries. They deliver impressive capacity and outstanding stability during charge-discharge cycles, making them a superior choice for enhancing battery performance and longevity. Jiménez et al. (Jiménez et al., 2017) have introduced a groundbreaking form of conducting, n-doped PANi, achieved through deprotonation in a lithium electrolyte. This innovative material delivers an exceptional blend of capacity, stability, coulombic efficiency, and both energy and power density, making it a highly promising positive electrode material for lithium-ion batteries. Its remarkable performance positions it at the forefront of battery technology advancements.

6.2 Supercapacitors

SCs are a groundbreaking technology in electrochemical energy storage, valued for their high power density, long lifespan, and rapid charge/discharge capabilities. They effectively bridge the gap between traditional dielectric capacitors, with their excellent power output, and batteries or fuel cells, which excel in energy storage (Largeot et al., 2008). However, SC technology faces significant challenges, including low energy density and high production costs, which hinder broader adoption. The energy storage capacity of supercapacitors is closely linked to the materials used for their electrodes (Wang et al., 2012). Notably, SCs that utilize redox-active materials enhance performance by not only accumulating charge at the electrical double layer but also providing pseudocapacitance through reversible redox reactions. This results in significantly higher specific capacitance and energy density compared to conventional electrical double-layer capacitors (EDLCs) (Esser et al., 2021). SCs can be classified into EDLCs, pseudocapacitors, and hybrid-type capacitors with battery-like behaviors, based on their charge storage mechanisms.

In recent decades, conventional conducting polymers, such as PANi, PPy, and polythiophene (PTh), have become crucial for redox polymers in SC electrodes due to their high conductivity, electrochemical activity, and easy synthesis. However, they face significant challenges, including incomplete doping, voltage fluctuations, and reduced lifespan caused by volume changes during electrochemical processes (Meng et al., 2017). In response, researchers are developing new redox polymers with more stable structures, advancing the potential of organic SC technology. These polymers' electrochemical responses are classified into three main types: p-doping, n-doping, and bipolar doping, paving the way for further innovation in the field (Lee et al., 2018).

Redox polymer-based supercapacitor devices consist of the following essential components: current collector, electrode, electrolyte, and separator, typically arranged in a 'sandwiched' architecture. These polymers are crucial as either redox-active materials providing electrochemical activity (Zhang et al., 2021) or as binders (Sandu et al., 2017) in the electrodes. This work highlights the latest developments in redox polymers as active electrode materials for supercapacitors. Traditionally used in powder form, these polymers require additives like adhesives and conductive carbon, which can diminish their capacitive performance. Recent innovations have introduced free-standing redox polymer-based electrodes, such as hydrogels and aerogels (Vivian Rachel Feig et al., 2019), that operate effectively without additives. This advancement is particularly relevant given the rise of portable and wearable electronics, where flexible supercapacitors are becoming the preferred power source, needing reliable performance under bending, folding, and stretching. Thanks to their excellent mechanical flexibility, processability, cost-effectiveness, and environmental friendliness, redox polymers are ideal for flexible SC applications. They have already demonstrated their potential in various flexible SC devices, including all-solid-state (Xu et al., 2017; Zhang et al., 2016) and paper-based designs (Zhang et al., 2015) showcasing their transformative role in energy storage technologies.

6.3 Thermoelectrics

Thermoelectric (TE) technology, which converts waste heat into electrical energy in a quiet and environmentally benign manner, is attracting unprecedented interest (Beretta et al., 2019). The distinct electrical and electronic characteristics of conjugated polymers position them as potential alternatives in metal-free organic electronics, including solar cells, photodiodes, field-effect transistors, and TEs. TE materials have gained new interest over the years because of the rising energy demands and the growing focus on renewable energy, along with remote and distributed applications. They allow for energy collection from various heat sources and, notably, from the abundant availability of solar thermal energy and waste heat present in our everyday experiences (Elsheikh et al., 2014; Heremans, 2014). Recent advancements in high-mobility conjugated polymers have generated interest in their use in TE applications for low-temperature heat (Zhang and Di, 2020). The variety in their chemical structures, ability to be processed in solution, lightweight design, favorable mechanical characteristics, and environmentally friendly qualities, along with their naturally low thermal conductivity, render them potential candidates for TE materials.

Although the advancement of conjugated polymers for TE applications has accelerated significantly in recent years, the focus of research in this field primarily revolves around p-type materials (Jia and Lei, 2019). The gradual advancement in n-type conjugated polymer TEs can be attributed to various factors. To begin with, when compared to p-type conjugated polymers, there is a smaller number of n-type conjugated polymers that exhibit high performance, as conventional aromatic/π moieties tend to donate electrons, resulting in enhanced delocalization of the highest occupied molecular orbital (HOMO) instead of the lowest unoccupied molecular orbital (LUMO). Secondly, to obtain n-type characteristics, the LUMO of the conjugated polymer must be low enough to enable effective electron injection and extraction, as well as stable electron transport (Sun et al., 2020). Finally, the restricted selection of n-dopants and the requirement for an inert atmosphere when working with n-doped conjugated polymers create both physical and mental challenges for researchers in this field. Therefore, it is clear that designing air-stable n-doped conjugated polymers necessitates the consideration of kinetic stability. That being noted, a TE device that is properly encapsulated can block the entry of oxygen and water, thereby guaranteeing the stability of the device (Tam and Xu, 2021).

7. Conclusion

Conductive polymers have emerged as vital components in a myriad of scientific and technological sectors, thanks to their remarkable ability to conduct electricity.

Their potential for developing innovative smart materials through targeted electrochemical modifications is truly exciting. Redox polymers, renowned for their electroactivity and remarkable capacity to undergo oxidation and reduction reactions, are indispensable for advancements in energy conversion technologies. Conjugated polymers, distinguished by their alternating single and double-bond structures, exhibit extraordinary properties that position them at the forefront of biomedical applications and energy storage solutions. This chapter has rigorously explored the synthesis and characterization techniques of conjugated redox polymers while highlighting their transformative applications, particularly in medicine and energy storage. By thoroughly understanding and strategically leveraging these unique properties, we can unlock the immense potential of these polymers, driving significant advancements in these critical fields.

References

Abidian, M.R., Daneshvar, E.D., Egeland, B.M., Kipke, D.R., Cederna, P.S. and Urbanchek, M.G. 2012. Hybrid conducting polymer–hydrogel conduits for axonal growth and neural tissue engineering. Advanced Healthcare Materials 2012 Nov; 1(6): 762–767. DOI: 10.1002/adhm.201200182. PMID: 23184828.

Ajjan, F., Casado, N., Rębiś, T., Elfwing, A., Solin, N., Mecerreyes, D. et al. 2016. High performance PEDOT/lignin biopolymer composites for electrochemical supercapacitors. Journal of Materials Chemistry A 4(5): 1838–1847.

Beretta, D., Neophytou, N., Hodges, J.M., Kanatzidis, M.G., Narducci, D., Martin-Gonzalez, M. et al. 2019. Thermoelectrics: From history, a window to the future. Materials Science and Engineering: R: Reports 138: 100501.

Biglari, N. and Zare, E.N. 2024. Conjugated polymer-based composite scaffolds for tissue engineering and regenerative medicine. Alexandria Engineering Journal 87: 277–299.

Boeva, Z.A. and Sergeyev, V.G. 2014. Polyaniline: Synthesis, properties, and application. Polymer Science Series C 56(1): 144–153. doi:10.1134/S1811238214010032.

Boyer, C., Teo, J., Phillips, P., Erlich, R.B., Sagnella, S., Sharbeen, G. et al. 2013. Effective delivery of siRNA into cancer cells and tumors using well-defined biodegradable cationic star polymers. Molecular Pharmaceutics 10(6): 2435–2444.

Breukers, R., Gilmore, K., Kita, M., Wagner, K., Higgins, M., Moulton, S. et al. 2010. Creating conductive structures for cell growth: Growth and alignment of myogenic cell types on polythiophenes. Journal of Biomedical Materials Research Part A 95(1): 256–268.

Broda, C.R., Lee, J.Y., Sirivisoot, S., Schmidt, C.E. and Harrison, B.S. 2011. A chemically polymerized electrically conducting composite of polypyrrole nanoparticles and polyurethane for tissue engineering. J. Biomed. Mater. Res. 98A(4): 509–516. doi:https://doi.org/10.1002/jbm.a.33128.

Casado, N., Hernandez, G., Sardon, H. and Mecerreyes, D. 2016. Current trends in redox polymers for energy and medicine. Progress in Polymer Science 52: 107–135.

Chauhan, N.P.S., Mozafari, M., Chundawat, N.S., Meghwal, K., Ameta, R. and Ameta, S.C. 2016. High-performance supercapacitors based on polyaniline–graphene nanocomposites: Some approaches, challenges and opportunities. Journal of Industrial and Engineering Chemistry 36: 13–29.

Chauhan, N.P.S., Jadoun, S., Rathore, B.S., Barani, M. and Zarrintaj, P. 2021. Redox polymers for capacitive energy storage applications. Journal of Energy Storage 43: 103218.

Chen, J., Dong, R., Ge, J., Guo, B. and Ma, P.X. 2015. Biocompatible, biodegradable, and electroactive polyurethane-urea elastomers with tunable hydrophilicity for skeletal muscle tissue engineering. ACS Applied Materials & Interfaces 7(51): 28273–28285. doi:10.1021/acsami.5b10829.

Chen, J., Yu, M., Guo, B., Ma, P.X. and Yin, Z. 2018. Conductive nanofibrous composite scaffolds based on *in-situ* formed polyaniline nanoparticle and polylactide for bone regeneration. Journal of Colloid and Interface Science 514: 517–527. doi:https://doi.org/10.1016/j.jcis.2017.12.062.

Chen, M.-C., Sun, Y.-C. and Chen, Y.-H. 2013. Electrically conductive nanofibers with highly oriented structures and their potential application in skeletal muscle tissue engineering. Acta Biomaterialia 9(3): 5562–5572. doi:https://doi.org/10.1016/j.actbio.2012.10.024.

Chen, Z., Zhang, H., He, X., Fan, G., Li, X., He, Z. et al. 2021. Fabrication of cellulosic paper containing zeolitic imidazolate framework and its application in removal of anionic dye from aqueous solution. BioResources 16(2).

Chiang, C.K., Fincher Jr, C., Park, Y.W., Heeger, A.J., Shirakawa, H., Louis, E.J. et al. 1977. Electrical conductivity in doped polyacetylene. Physical Review Letters 39(17): 1098.

Chundawat, N.S., Pande, N., Sargazi, G., Gholipourmalekabadi, M. and Chauhan, N.P.S. 2020a. Structure-properties relationship for energy storage redox polymers: a review. Journal of Polymer Engineering 40(5): 373–393.

Chundawat, N.S., Pande, N., Sargazi, G., Gholipourmalekabadi, M. and Chauhan, N.P.S. 2020b. Structure-properties relationship for energy storage redox polymers: a review. Journal of Polymer Engineering 40(5): 373–393. doi:doi:10.1515/polyeng-2019-0395.

Deng, Z., Guo, Y., Zhao, X., Li, L., Dong, R., Guo, B. et al. 2016. Stretchable degradable and electroactive shape memory copolymers with tunable recovery temperature enhance myogenic differentiation. Acta Biomaterialia 46: 234–244. doi:https://doi.org/10.1016/j.actbio.2016.09.019.

Deraedt, C., Rapakousiou, A., Wang, Y., Salmon, L., Bousquet, M. and Astruc, D. 2014. Multifunctional redox polymers: electrochrome, polyelectrolyte, sensor, electrode modifier, nanoparticle stabilizer, and catalyst template. Angewandte Chemie International Edition 53(32): 8445–8449.

Dong, R., Zhao, X., Guo, B. and Ma, P.X. 2016. Self-healing conductive injectable hydrogels with antibacterial activity as cell delivery carrier for cardiac cell therapy. ACS Applied Materials & Interfaces 8(27): 17138–17150. doi:10.1021/acsami.6b04911.

Dubey, A.K., Gupta, S.D. and Basu, B. 2011. Optimization of electrical stimulation parameters for enhanced cell proliferation on biomaterial surfaces. Journal of Biomedical Materials Research Part B: Applied Biomaterials 98(1): 18–29.

Elsheikh, M.H., Shnawah, D.A., Sabri, M.F.M., Said, S.B.M., Hassan, M.H., Bashir, M.B.A. et al. 2014. A review on thermoelectric renewable energy: Principle parameters that affect their performance. Renewable and Sustainable Energy Reviews 30: 337–355.

Esser, B., Dolhem, F., Becuwe, M., Poizot, P., Vlad, A. and Brandell, D. 2021. A perspective on organic electrode materials and technologies for next generation batteries. Journal of Power Sources 482: 228814. doi:https://doi.org/10.1016/j.jpowsour.2020.228814.

Feig, V.R., Tran, H., Lee, M. and Bao, Z. 2018. Mechanically tunable conductive interpenetrating network hydrogels that mimic the elastic moduli of biological tissue. Nature Communications 9(1): 2740. doi:10.1038/s41467-018-05222-4.

Feig, V.R., Tran, H., Lee, M., Liu, K., Huang, Z., Beker, L. et al. 2019. An electrochemical gelation method for patterning conductive PEDOT: PSS hydrogels. Advanced Materials 31(39): 1902869.

Foroughi, J., Spinks, G.M. and Wallace, G.G. 2011. A reactive wet spinning approach to polypyrrole fibres. Journal of Materials Chemistry 21(17): 6421–6426.

Gelmi, A., Cieslar-Pobuda, A., de Muinck, E., Los, M., Rafat, M. and Jager, E.W. 2016. Direct mechanical stimulation of stem cells: a beating electromechanically active scaffold for cardiac tissue engineering. Advanced Healthcare Materials 5(12): 1471–1480.

Genene, Z., Xia, Z., Yang, G., Mammo, W. and Wang, E. 2024. Recent advances in the synthesis of conjugated polymers for supercapacitors. Advanced Materials Technologies 9(9): 2300167.

Ghasemi-Mobarakeh, L., Prabhakaran, M.P., Morshed, M., Nasr-Esfahani, M.H. and Ramakrishna, S. 2009. Electrical stimulation of nerve cells using conductive nanofibrous scaffolds for nerve tissue engineering. Tissue Engineering Part A 15(11): 3605–3619. doi:10.1089/ten.tea.2008.0689.

Gomez, N. and Schmidt, C.E. 2007. Nerve growth factor-immobilized polypyrrole: Bioactive electrically conducting polymer for enhanced neurite extension. J. Biomed. Mater. Res. 81A(1): 135–149. doi:https://doi.org/10.1002/jbm.a.31047.

Harris, A.R., Morgan, S.J., Chen, J., Kapsa, R.M., Wallace, G.G. and Paolini, A.G. 2012. Conducting polymer coated neural recording electrodes. Journal of Neural Engineering 10(1): 016004.

Heinze, J., Frontana-Uribe, B.A. and Ludwigs, S. 2010. Electrochemistry of conducting polymers-persistent models and new concepts. Chemical Reviews 110(8): 4724–4771.

Heremans, J.P. 2014. The ugly duckling. Nature 508(7496): 327–328.

Hsiao, C.-W., Bai, M.-Y., Chang, Y., Chung, M.-F., Lee, T.-Y., Wu, C.-T. et al. 2013. Electrical coupling of isolated cardiomyocyte clusters grown on aligned conductive nanofibrous meshes for their synchronized beating. Biomaterials 34(4): 1063–1072. doi:https://doi.org/10.1016/j.biomaterials.2012.10.065.

Huang, Y., Li, H., Wang, Z., Zhu, M., Pei, Z., Xue, Q. et al. 2016. Nanostructured Polypyrrole as a flexible electrode material of supercapacitor. Nano Energy 22: 422–438. doi:https://doi.org/10.1016/j.nanoen.2016.02.047.

Huo, M., Yuan, J., Tao, L. and Wei, Y. 2014. Redox-responsive polymers for drug delivery: from molecular design to applications. Polymer Chemistry 5(5): 1519–1528. doi:10.1039/C3PY01192E.

Iandolo, D., Ravichandran, A., Liu, X., Wen, F., Chan, J.K.Y., Berggren, M. et al. 2016. Development and characterization of organic electronic scaffolds for bone tissue engineering. Adv. Healthcare Mater. 5(12): 1505–1512. doi:https://doi.org/10.1002/adhm.201500874.

Inal, S., Rivnay, J., Suiu, A.-O., Malliaras, G.G. and McCulloch, I. 2018. Conjugated polymers in bioelectronics. Accounts of Chemical Research 51(6): 1368–1376.

Jadoun, S., Chauhan, N.P.S., Chinnam, S., Aepuru, R., Sathish, M., Chundawat, N.S. et al. 2023. A short review on conducting polymer nanocomposites. Biomedical Materials & Devices 1(1): 351–365.

Jadoun, S., Yáñez, J., Mansilla, H.D., Riaz, U. and Chauhan, N.P.S. 2022. Conducting polymers/zinc oxide-based photocatalysts for environmental remediation: a review. Environmental Chemistry Letters 20(3): 2063–2083.

Jean-Rémi, P., François, G., Terence, B.J., Serge, B. and Mario, L. 2016. Direct (Hetero) arylation Polymerization: Simplicity for Conjugated Polymer Synthesis.

Jia, H. and Lei, T. 2019. Emerging research directions for n-type conjugated polymers. Journal of Materials Chemistry C 7(41): 12809–12821.

Jia, Z., Liu, J., Boyer, C., Davis, T.P. and Bulmus, V. 2009. Functional disulfide-stabilized polymer–protein particles. Biomacromolecules 10(12): 3253–3258.

Jiménez, P., Levillain, E., Alévêque, O., Guyomard, D., Lestriez, B. and Gaubicher, J. 2017. Lithium n-doped polyaniline as a high-performance electroactive material for rechargeable batteries. Angewandte Chemie International Edition 56(6): 1553–1556.

Jun, I., Jeong, S. and Shin, H. 2009. The stimulation of myoblast differentiation by electrically conductive sub-micron fibers. Biomaterials 30(11): 2038–2047. doi:https://doi.org/10.1016/j.biomaterials.2008.12.063.

Kai, D., Prabhakaran, M.P., Jin, G. and Ramakrishna, S. 2011. Polypyrrole-contained electrospun conductive nanofibrous membranes for cardiac tissue engineering. J. Biomed. Mater. Res. 99A(3): 376–385. doi:https://doi.org/10.1002/jbm.a.33200.

KarbalaeiMahdi, A., Moridi, K. and Ghollasi, M. 2022. Evaluation of osteogenic differentiation of human mesenchymal stem cells (hMSCs) on random and aligned polycaprolactone-polyaniline-gelatin scaffolds. BioImpacts: BI 13(2): 123.

Kausar, A. 2021. Conducting Polymer-based Nanocomposites: Fundamentals and Applications: Elsevier.

Kim, J., Kim, J.H. and Ariga, K. 2017. Redox-active polymers for energy storage nanoarchitectonics. Joule 1(4): 739–768.

Kotwal, A. and Schmidt, C.E. 2001. Electrical stimulation alters protein adsorption and nerve cell interactions with electrically conducting biomaterials. Biomaterials 22(10): 1055–1064. doi:https://doi.org/10.1016/S0142-9612(00)00344-6.

Larcher, D. and Tarascon, J.-M. 2015. Towards greener and more sustainable batteries for electrical energy storage. Nature Chemistry 7(1): 19–29.

Largeot, C., Portet, C., Chmiola, J., Taberna, P.-L., Gogotsi, Y. and Simon, P. 2008. Relation between the ion size and pore size for an electric double-layer capacitor. Journal of the American Chemical Society 130(9): 2730–2731. doi:10.1021/ja7106178.

Lee, S., Kwon, G., Ku, K., Yoon, K., Jung, S.K., Lim, H.D. et al. 2018. Recent progress in organic electrodes for Li and Na rechargeable batteries. Advanced Materials 30(42): 1704682.

Lewis, N.S. 2016. Research opportunities to advance solar energy utilization. Science 351(6271): aad1920.

Li, Z., Peng, M., Zhou, X., Shin, K., Tunmee, S., Zhang, X. et al. 2021. *In situ* chemical lithiation transforms diamond-like carbon into an ultrastrong ion conductor for dendrite-free lithium-metal anodes. Adv. Mater. 33(37): 2100793.

Liu, X., Yue, Z., Higgins, M.J. and Wallace, G.G. 2011. Conducting polymers with immobilised fibrillar collagen for enhanced neural interfacing. Biomaterials 32(30): 7309–7317. doi:https://doi.org/10.1016/j.biomaterials.2011.06.047.

Liu, Y., Gao, S., Zhang, X., Xin, J.H. and Zhang, C. 2023. Probing the nature of charge carriers in one-dimensional conjugated polymers: a review of the theoretical models, experimental trends, and thermoelectric applications. Journal of Materials Chemistry C 11(1): 12–47.

Malik, A.H., Habib, F., Qazi, M.J., Ganayee, M.A., Ahmad, Z. and Yatoo, M.A. 2023. A short review article on conjugated polymers. Journal of Polymer Research 30(3): 115.

Marriam, I., Wang, Y. and Tebyetekerwa, M. 2020. Polyindole batteries and supercapacitors. Energy Storage Materials 33: 336–359. doi:https://doi.org/10.1016/j.ensm.2020.08.010.

Mawad, D., Mansfield, C., Lauto, A., Perbellini, F., Nelson, G.W., Tonkin, J. et al. 2016. A conducting polymer with enhanced electronic stability applied in cardiac models. Science Advances 2(11): e1601007. doi:doi:10.1126/sciadv.1601007.

Meng, Q., Cai, K., Chen, Y. and Chen, L. 2017. Research progress on conducting polymer based supercapacitor electrode materials. Nano Energy 36: 268–285. doi:https://doi.org/10.1016/j.nanoen.2017.04.040.

Meng, S., Rouabhia, M., Shi, G. and Zhang, Z. 2008. Heparin dopant increases the electrical stability, cell adhesion, and growth of conducting polypyrrole/poly(L,L-lactide) composites. Journal of Biomedical Materials Research Part A: An Official Journal of The Society for Biomaterials, The Japanese Society for Biomaterials, and The Australian Society for Biomaterials and the Korean Society for Biomaterials 87A(2): 332–344. doi:https://doi.org/10.1002/jbm.a.31735.

Meng, S., Rouabhia, M. and Zhang, Z. 2013. Electrical stimulation modulates osteoblast proliferation and bone protein production through heparin-bioactivated conductive scaffolds. Bioelectromagnetics 34(3): 189–199. doi:https://doi.org/10.1002/bem.21766.

Mihic, A., Cui, Z., Wu, J., Vlacic, G., Miyagi, Y., Li, S.-H. et al. 2015. A conductive polymer hydrogel supports cell electrical signaling and improves cardiac function after implantation into myocardial infarct. Circulation 132(8): 772–784. doi:doi:10.1161/CIRCULATIONAHA.114.014937.

Molino, B.Z., Fukuda, J., Molino, P.J. and Wallace, G.G. 2021. Redox polymers for tissue engineering. Frontiers in Medical Technology 3: 669763.

Molino, P.J., Garcia, L., Stewart, E.M., Lamaze, M., Zhang, B., Harris, A.R. et al. 2018. PEDOT doped with algal, mammalian and synthetic dopants: polymer properties, protein and cell interactions, and influence of electrical stimulation on neuronal cell differentiation. Biomaterials Science 6(5): 1250–1261.

Moutsatsou, P., Coopman, K. and Georgiadou, S. 2017. Biocompatibility assessment of conducting PANi/chitosan nanofibers for wound healing applications. Polymers 9(12): 687.

Mozafari, M. and Chauhan, N.P.S. 2019. Fundamentals and emerging applications of polyaniline.

Pande, N., Jambhale, A., Purohit, V., Jaspal, D., Ambekar, J. and Chauhan, N.P.S. 2021. Poly (o-toluidine)-Li nanocomposite: facile preparation and utilization in energy storage devices. Polymer-Plastics Technology and Materials 60(2): 208–216.

Parashar, K., Prajapati, D., McIntyre, R. and Kandasubramanian, B. 2020. Advancements in biological neural interfaces using conducting polymers: a review. Industrial & Engineering Chemistry Research 59(21): 9707–9718. doi:10.1021/acs.iecr.0c00174.

Park, M.-S., Kim, J., Kim, K.J., Lee, J.-W., Kim, J.H. and Yamauchi, Y. 2015. Porous nanoarchitectures of spinel-type transition metal oxides for electrochemical energy storage systems. Physical Chemistry Chemical Physics 17(46): 30963–30977. doi:10.1039/C5CP05936D.

Prabhakaran, M.P., Ghasemi-Mobarakeh, L., Jin, G. and Ramakrishna, S. 2011. Electrospun conducting polymer nanofibers and electrical stimulation of nerve stem cells. Journal of Bioscience and Bioengineering 112(5): 501–507.

Qiu, Z., Hammer, B.A. and Müllen, K. 2020. Conjugated polymers–Problems and promises. Progress in Polymer Science 100: 101179.

Ramya, R., Sivasubramanian, R. and Sangaranarayanan, M. 2013. Conducting polymers-based electrochemical supercapacitors—Progress and prospects. Electrochimica Acta 101: 109–129.

Richardson, R.T., Thompson, B., Moulton, S., Newbold, C., Lum, M.G., Cameron, A. et al. 2007. The effect of polypyrrole with incorporated neurotrophin-3 on the promotion of neurite outgrowth from auditory neurons. Biomaterials 28(3): 513–523.

Sandu, G., Ernould, B., Rolland, J., Cheminet, N., Brassinne, J., Das, P.R. et al. 2017. Mechanochemical synthesis of PEDOT: PSS hydrogels for aqueous formulation of Li-ion battery electrodes. ACS Applied Materials & Interfaces 9(40): 34865–34874.

Sarkar, S. and Levi-Polyachenko, N. 2020. Conjugated polymer nano-systems for hyperthermia, imaging and drug delivery. Advanced Drug Delivery Reviews 163: 40–64.

Schon, T.B., McAllister, B.T., Li, P.-F. and Seferos, D.S. 2016. The rise of organic electrode materials for energy storage. Chemical Society Reviews 45(22): 6345–6404.

Sharifabad, A.H., Ghanbari, R., Saeb, M.R., Zare, E.N., Rossi, F., Salami-Kalajahi, M. et al. 2025. 3D Engineered scaffolds of conjugated polymers/metal organic frameworks for biomedical applications. International Materials Reviews 09506608241299174.

Shi, G., Rouabhia, M., Meng, S. and Zhang, Z. 2008. Electrical stimulation enhances viability of human cutaneous fibroblasts on conductive biodegradable substrates. Journal of Biomedical Materials Research Part A: An Official Journal of The Society for Biomaterials, The Japanese Society for Biomaterials, and The Australian Society for Biomaterials and the Korean Society for Biomaterials 84A(4): 1026–1037. doi:https://doi.org/10.1002/jbm.a.31337.

Shi, Z., Gao, H., Feng, J., Ding, B., Cao, X., Kuga, S. et al. 2014. *In situ* synthesis of robust conductive cellulose/polypyrrole composite aerogels and their potential application in nerve regeneration. Angew Chem Int Ed Engl. 53(21): 5380–5384. doi:10.1002/anie.201402751.

Shirakawa, H., Louis, E.J., MacDiarmid, A.G., Chiang, C.K. and Heeger, A.J. 1977. Synthesis of electrically conducting organic polymers: halogen derivatives of polyacetylene, (CH) x. Journal of the Chemical Society, Chemical Communications (16): 578–580.

Song, N., Liu, W., Tu, Q., Liu, R., Zhang, Y. and Wang, J. 2011. Preparation and *in vitro* properties of redox-responsive polymeric nanoparticles for paclitaxel delivery. Colloids and Surfaces B: Biointerfaces 87(2): 454–463. doi:https://doi.org/10.1016/j.colsurfb.2011.06.009.

Stamenkovic, V.R., Strmcnik, D., Lopes, P.P. and Markovic, N.M. 2017. Energy and fuels from electrochemical interfaces. Nature Materials 16(1): 57–69.

Stewart, E.M., Fabretto, M., Mueller, M., Molino, P.J., Griesser, H.J., Short, R.D. et al. 2013. Cell attachment and proliferation on high conductivity PEDOT–glycol composites produced by vapour phase polymerisation. Biomaterials Science 1(4): 368–378.

Sun, B., Zhou, Z., Li, D., Wu, T., Zheng, H., Liu, J. et al. 2019. Polypyrrole-coated poly(l-lactic acid-co-ε-caprolactone)/silk fibroin nanofibrous nerve guidance conduit induced nerve regeneration in rat. Materials Science and Engineering: C 94: 190–199. doi:https://doi.org/10.1016/j.msec.2018.09.021.

Sun, H., Guo, X. and Facchetti, A. 2020. High-performance n-type polymer semiconductors: applications, recent development, and challenges. Chem. 6(6): 1310–1326.

Sun, L., Li, C., Zhang, C., Liang, T. and Zhao, Z. 2019. The strain transfer mechanism of fiber bragg grating sensor for extra large strain monitoring. Sensors 19(8): 1851.

Tam, T.L.D. and Xu, J. 2021. Strategies and concepts in n-doped conjugated polymer thermoelectrics. Journal of Materials Chemistry A 9(9): 5149–5163.

Tandon, N., Goh, B., Marsano, A., Chao, P.-H.G., Montouri-Sorrentino, C., Gimble, J. et al. 2009. Alignment and elongation of human adipose-derived stem cells in response to direct-current electrical stimulation. Paper Presented at the 2009 Annual International Conference of the IEEE Engineering in Medicine and Biology Society.

Thompson, B.C., Moulton, S.E., Richardson, R.T. and Wallace, G.G. 2011. Effect of the dopant anion in polypyrrole on nerve growth and release of a neurotrophic protein. Biomaterials 32(15): 3822–3831.

Thompson, B.C., Richardson, R.T., Moulton, S.E., Evans, A.J., O'Leary, S., Clark, G.M. et al. 2010. Conducting polymers, dual neurotrophins and pulsed electrical stimulation—dramatic effects on neurite outgrowth. Journal of Controlled Release 141(2): 161–167.

Urban, M.W. 2011. Handbook of Stimuli-Responsive Materials. John Wiley & Sons.

Vijayakumar, V., Anothumakkool, B., Kurungot, S., Winter, M. and Nair, J.R. 2021. *In situ* polymerization process: an essential design tool for lithium polymer batteries. Energy & Environmental Science 14(5): 2708–2788.

Wallace, G.G., Teasdale, P.R., Spinks, G.M. and Kane-Maguire, L.A. 2008. Conductive Electroactive Polymers: Intelligent Polymer Systems: CRC press.

Wang, G., Zhang, L. and Zhang, J. 2012. A review of electrode materials for electrochemical supercapacitors. Chemical Society Reviews 41(2): 797–828. doi:10.1039/C1CS15060J.

Wang, L., Wu, Y., Hu, T., Guo, B. and Ma, P.X. 2017. Electrospun conductive nanofibrous scaffolds for engineering cardiac tissue and 3D bioactuators. Acta Biomaterialia 59: 68–81. doi:https://doi.org/10.1016/j.actbio.2017.06.036.

Wang, S., Sun, C., Guan, S., Li, W., Xu, J., Ge, D. et al. 2017. Chitosan/gelatin porous scaffolds assembled with conductive poly(3,4-ethylenedioxythiophene) nanoparticles for neural tissue engineering. Journal of Materials Chemistry B 5(24): 4774–4788. doi:10.1039/C7TB00608J.

Wang, Y., Chu, X., Zhang, H., Yan, C., Tian, G., Yang, W. et al. 2023. Hyper-conjugated polyaniline delivering extraordinary electrical and electrochemical properties in supercapacitors. Applied Surface Science 628: 157350. doi:https://doi.org/10.1016/j.apsusc.2023.157350.

Wang, Y., Rouabhia, M. and Zhang, Z. 2016. Pulsed electrical stimulation benefits wound healing by activating skin fibroblasts through the TGFβ1/ERK/NF-κB axis. Biochimica et Biophysica Acta (BBA) - General Subjects 1860(7): 1551–1559. doi:https://doi.org/10.1016/j.bbagen.2016.03.023.

Wu, J., Wang, J., Huang, X. and Bai, H. 2018. A self-assembly route to porous polyaniline/reduced graphene oxide composite materials with molecular-level uniformity for high-performance supercapacitors. Energy & Environmental Science 11(5): 1280–1286.

Xia, L., Wei, Z. and Wan, M. 2010. Conducting polymer nanostructures and their application in biosensors. Journal of Colloid and Interface Science 341(1): 1–11. doi:https://doi.org/10.1016/j.jcis.2009.09.029.

Xie, J., Zhao, C.e., Lin, Z.q., Gu, P.y. and Zhang, Q. 2016. Nanostructured conjugated polymers for energy-related applications beyond solar cells. Chemistry–An Asian Journal 11(10): 1489–1511.

Xie, M., Wang, L., Ge, J., Guo, B. and Ma, P.X. 2015. Strong electroactive biodegradable shape memory polymer networks based on star-shaped polylactide and aniline trimer for bone tissue engineering. ACS Applied Materials & Interfaces 7(12): 6772–6781. doi:10.1021/acsami.5b00191.

Xu, C., Guan, S., Wang, S., Gong, W., Liu, T., Ma, X. et al. 2018. Biodegradable and electroconductive poly(3,4-ethylenedioxythiophene)/carboxymethyl chitosan hydrogels for neural tissue engineering. Materials Science and Engineering: C 84: 32–43. doi:https://doi.org/10.1016/j.msec.2017.11.032.

Xu, X., Tang, J., Qian, H., Hou, S., Bando, Y., Hossain, M.S.A. et al. 2017. Three-dimensional networked metal–organic frameworks with conductive polypyrrole tubes for flexible supercapacitors. ACS Applied Materials & Interfaces 9(44): 38737–38744.

Yiğit, D. and Güllü, M. 2017. N-Substituted poly (3, 6-dithienylcarbazole) derivatives: a new class of redox-active electrode materials for high-performance flexible solid-state pseudocapacitors. Journal of Materials Chemistry A 5(2): 609–618.

Yuan, M. and Minteer, S.D. 2019. Redox polymers in electrochemical systems: From methods of mediation to energy storage. Current Opinion in Electrochemistry 15: 1–6. doi:https://doi.org/10.1016/j.coelec.2019.03.003.

Zanjanizadeh Ezazi, N., Shahbazi, M.-A., Shatalin, Y.V., Nadal, E., Mäkilä, E., Salonen, J. et al. 2018. Conductive vancomycin-loaded mesoporous silica polypyrrole-based scaffolds for bone regeneration. International Journal of Pharmaceutics 536(1): 241–250. doi:https://doi.org/10.1016/j.ijpharm.2017.11.065.

Zhang, F. and Di, C.-a. 2020. Exploring thermoelectric materials from high mobility organic semiconductors. Chemistry of Materials 32(7): 2688–2702.

Zhang, L., Liu, Z., Cui, G. and Chen, L. 2015. Biomass-derived materials for electrochemical energy storages. Progress in Polymer Science 43: 136–164.

Zhang, Q., He, Y., Lin, G., Ma, X., Xiao, Z., Shi, D. et al. 2021. *In situ* growth of polyimide nanoarrays on conductive carbon supports for high-rate charge storage and long-lived metal-free cathodes. Journal of Materials Chemistry A 9(17): 10652–10660. doi:10.1039/D1TA00302J.

Zhang, S., Kumar, P., Nouas, A.S., Fontaine, L., Tang, H. and Cicoira, F. 2015. Solvent-induced changes in PEDOT: PSS films for organic electrochemical transistors. APL Materials 3(1).

Zhang, Y.-Z., Wang, Y., Cheng, T., Lai, W.-Y., Pang, H. and Huang, W. 2015. Flexible supercapacitors based on paper substrates: a new paradigm for low-cost energy storage. Chemical Society Reviews 44(15): 5181–5199. doi:10.1039/C5CS00174A.

Zhang, Y.Z., Cheng, T., Wang, Y., Lai, W.Y., Pang, H. and Huang, W. 2016. A simple approach to boost capacitance: flexible supercapacitors based on manganese oxides@ MOFs via chemically induced *in situ* self-transformation. Advanced Materials 28(26): 5242–5248.

Zhao, M., Biswas, A., Hu, B., Joo, K.-I., Wang, P., Gu, Z. et al. 2011. Redox-responsive nanocapsules for intracellular protein delivery. Biomaterials 32(22): 5223–5230. doi:https://doi.org/10.1016/j.biomaterials.2011.03.060.

Zhou, H., Peng, J., Duan, X., Yin, H., Huang, B., Zhou, C. et al. 2023. Redox-active polymers as robust electron-shuttle co-catalysts for fast Fe^{3+}/Fe^{2+} circulation and green fenton oxidation. Environmental Science & Technology 57(8): 3334–3344. doi:10.1021/acs.est.2c07447.

Chapter 3

Composite Polymer Electrolytes for Battery Applications

Nandha Gopal Balasubramaniyan,[1] *Panneerselvam Perumal,*[1,*]
Lajvanti Vyas,[2] *Kamal Jeet Yadav,*[2] *Gyan Singh,*[3]
Narendra Pal Singh Chauhan[2] and *Sapana Jadoun*[3]

1. Introduction

The rising demand for compact, high-performance, and sustainable technology, including laptops, wearables, tablets, and smartphones, necessitates advanced energy storage solutions. These devices necessitate energy storage devices (ESDs) that are reliable, smart, and easy to maintain. To support sustainable energy initiatives, electrochemical storage systems are indispensable. High-energy-density materials are critical for developing effective electrochemical capacitors (Abbasi et al., 2017; Chameh et al., 2021, 2020; Mehtab et al., 2019). As revealed in Fig. 1, ESD can be broadly classified into four types based on energy bases. The first type of battery (primary and secondary), which is quite widely used owing to its excellent performance in several applications, has achieved good acceptance. As battery technology advances, there is a strong focus on optimizing battery performance for various uses. A rechargeable battery is composed of a cathode, an anode, and

[1] Department of Chemistry, SRM Institute of Science and Technology, Kattankulathur, Chengalpattu, Tamil Nadu 603203, India.
[2] Department of Chemistry, Faculty of Science, Bhupal Nobles' University, Udaipur, Rajasthan, India.
[3] Sol-ARIS, Departamento de Química, Facultad de Ciencias, Universidad de Tarapacá, Avda. General Velásquez, 1775, Arica, Chile.
* Corresponding author: panneerp1@srmist.edu.in, panneerchem82@gmail.com

DOI: 10.1201/9781003546672-3

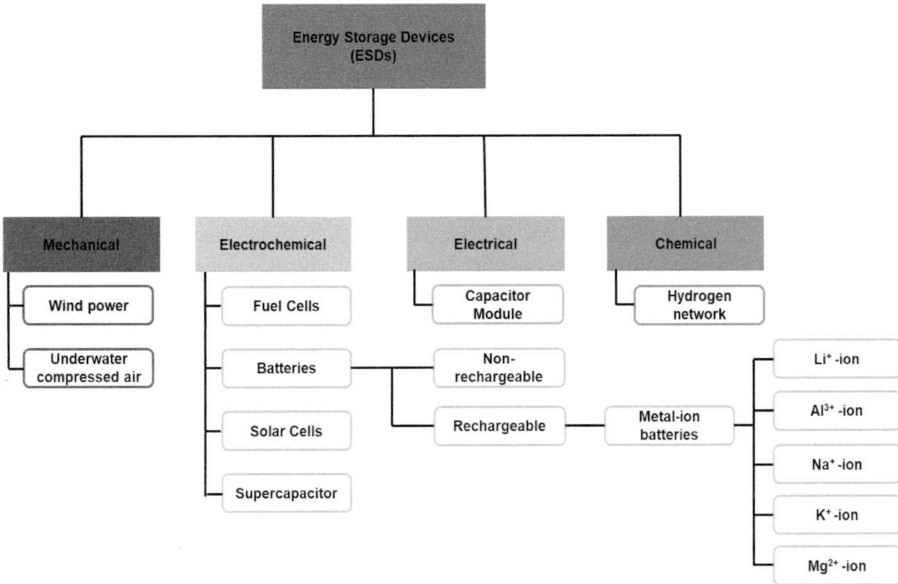

Fig. 1. A system for categorizing energy storage devices.

an electrolyte (Krishan and Suhag, 2019). Electrodes are key components in the battery's energy generation mechanism. The electrolyte, either solid or gel-type, facilitates ion movement between the cathode and anode, allowing the battery to discharge energy. The other variants consist of a membrane or a separator made from different materials. This ensures that a direct short circuit does not happen between the electrodes during operation. In reverse order, it works during recharging. The most researched metal-ion batteries include lithium, zinc, sodium, potassium, magnesium, etc. (Liang et al., 2020a).

Metal-ion batteries face several challenges, including small specific capacity, limited energy concentration, protection concerns, great ionic radii, hindered ion diffusion, and sluggish electrochemical kinetics. Charging and discharging rates become restricted because the system causes considerable volume changes due to cycling, which leads to fragile mechanical behavior. Hence, these materials must possess intrinsic physical and electrochemical attributes such as electronic and ionic conductivity, redox potential, capacity, and catalytic movement. As such, innovative architectures and chemical compositions are needed. Each system of batteries has these high-performance materials actively under development to overcome such challenges (Pan et al., 2021).

Inorganic materials identical to metal-organic frameworks (MOFs) and their polymer hybrids exhibit potential for advancing metal-ion battery technology. These materials offer advantages such as fire safety, short-circuit prevention, and high energy storage capacities. MOFs are crystalline porous materials made up of metal particles or clusters linked by organic ligands, resulting in diverse and orderly structures. These are currently under investigation for a range of applications including drug delivery, gas adsorption, and energy storage. Their porous assembly facilitates fast electrolyte diffusion and efficient ion dissemination. In the case of having customizable structures and components, MOFs are seen with strong frameworks that afford excellent electrochemical stability and high capacity through numerous electroactive sites. However, MOFs are rarely used as-is as battery electrode materials since they possess low electrical conduction. MOF compounds and materials resulting from MOFs, on the other hand, retain all the advantages that MOFs possess in structure, porosity, and composition but acquire enhanced conductivity due to the presence of functional components added to them. One general approach to mitigating the typically low conductivity of MOFs is through the addition of conductive matrices or functionalization with guest molecules to enhance their electrical performance (Xie et al., 2020). Polymers are a class of solid materials that may be molded, shaped, and adapted for a plethora of everyday technologies. Beyond providing a template for guiding structural design, polymers can be transformed into carbon materials that make otherwise fragile electrode materials stable. MOF composites can be created by combining MOFs with functional materials like carbon, polymers, metal nanoparticles, and functional molecules. The interaction between MOFs and these components significantly improves the electrochemical concert of MOF complexes. MOF-polymer hybrids are widely used in recent energy-loading technology as cathodes, anodes, and separator/electrolyte classifications owed to the synergistic properties of each material. Research indicates that polymers influences the evolution and properties of MOFs, while MOFs can also affect the structure of polymers (Costa et al., 2020).

2. Assembly of Polymer-MOF Composites

The use of polymer-MOF hybrids has gained increased interest in recently discovered high-performance applications. Polymers supply flexibility, toughness, and an appreciable level of processability, whereas MOFs come in the rigid, crystalline counterpart of such material. A combination of these two would enable someone to keep the desirable properties of the polymer or MOF, and also open up for the creation of complex hybrids with novel properties arising from unpredicted host-guest interactions. Polymers facilitate the shaping of MOFs into films and pellets. This combination has also ignited discoveries in materials and polymer chemistry by exploring the interactions between these components in the structure, assembly, and properties of their hybrid materials. Pure MOFs typically exhibit poor conductivity; therefore, researchers have engineered them into composite MOFs with improved characteristics (Zhang et al., 2016). Polymer-MOF composites can be prepared both inside and outside the MOF pores. The polymers can be chemically linked to the MOF surface or held in place by less stable physical contact with the host framework. This chapter delves into five wide-ranging approaches for the synthesis of polymer-MOF composites (Fig. 2) (Yang et al., 2021a).

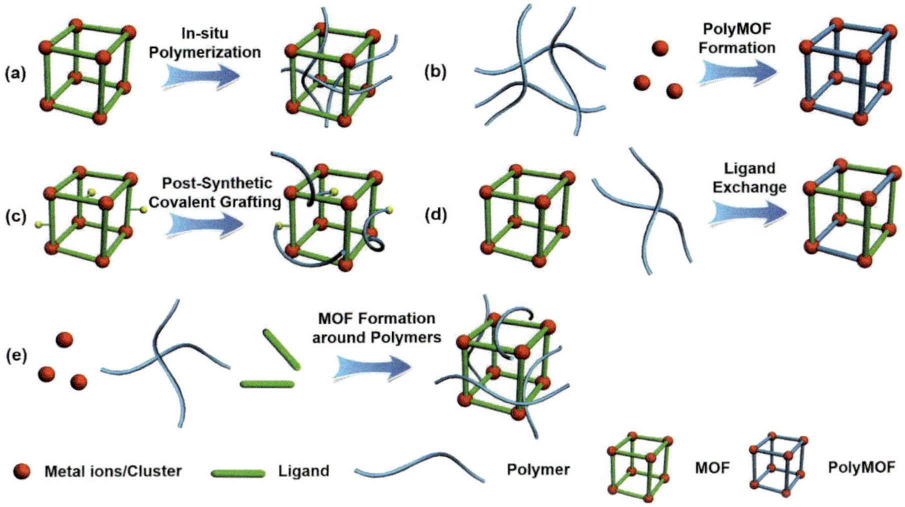

Fig. 2. The preparation of MOF composites often involves techniques such as (a) *in-situ* polymerization, (b) construction of MOFs from polymeric ligands (polyMOFs), (c) post-synthetic attaching of polymers onto MOF ligands, (d) post-synthetic polymeric ligand substitution, and (e) self-assembly of MOFs about polymers. This information was adapted with authorization from (Yang et al., 2021b) 2021, Elsevier.

2.1 In-Situ Polymerization of Monomers

First, monomers were polymerized within the pores of MOFs after diffusing in as gases or liquids. If polymerization starts, it may be promoted by light, heat, chemical initiators, or other specific reactions involving the monomers and the MOF structure. Polymers that are greater than the pore size are retained, which exerts a stabilizing influence with a "ship-in-a-bottle" type of effect. Intermolecular forces between the walls of the MOFs and guest polymers retain these latter polymers in the pores. To synthesize linear polymers with customizable repeating units, styrene can be polymerized within the pores of metal-organic frameworks (MOFs). This approach yields high-molecular-weight polystyrene with a slight polydispersity (Yang et al., 2021a). MOFs have pores that make the "living radical" classes used in styrene polymerization more stable, which makes them last longer. Researchers have found that MOFs can hold different vinyl polymers, such as polyvinyl acetate, polyacrylonitrile, polymethacrylic acid, and polymethyl methacrylate (PMMA). This extended period of the "living radical" species leads to the creation of highly monodispersed polymers, enabling the development of both cationic and anionic process. Specifically, cationic inventors offer control over the dispersity of three-dimensional polymers identical polyglucose synthesized within MOFs (Uemura et al., 2005). Additionally, anionic carboxylate groups on the MOF ligand have been demonstrated to initiate the polymerization of polyacetylene within a copper-based MOF. This is perhaps the first instance where MOFs were systematically applied as initiators. Prompted by these early observations, research efforts began to design protocols for synthesizing MOFs where active sites would drive polymerization reactions spontaneously, without external initiators. Based on this principle, several groups have employed *in-situ* polymerization. One such group is Ding et al. which polymerized 1,2-diethynylbenzene (DEB) directly within MOFs, thereby forming hydrophobic cavities in the open channels to enhance the CO_2 volume and the CO_2/N_2 discernment. In the article, DEB was first absorbed and captured into the MOF-5 channels and then animated up to high temperatures. After a Bergman cyclization reaction followed by radical polymerization, PN was synthesized in the channels. Such composite is known as PN@MOF-5.Although *in-situ* polymerization reactions offer convenience, they come with certain limitations. Controlling the process can be challenging, and polymerization on the MOF surface may result in pore blockage. Additionally, initiators and other reagents used for *in-situ* polymerization must be compatible with the MOF; otherwise, reactive initiators might lead to MOF decomposition (Ding et al., 2016; Gamage et al., 2016).

2.2 Construction of MOFs from Polymeric Ligands (polyMOFs)

In 2015, the term "polyMOFs" was coined for a new concept in generating polymer-MOF composites directly, through the inclusion of pre-synthesized polymeric ligands into the structures of MOF (Zhang et al., 2015). PolyMOFs thus retained the porous and crystalline character of MOFs, while possessing chemical firmness and processability offered by polymers. They produced the initial polymeric ligand in Cohen's laboratory through a reaction of diacyl dichloride of different alkyl spacers with 2-amino-1,4-benzenedicarboxylic acid, which when heated with the respective Zn^{2+} salt provided isoreticular variations of IRMOF-1. Following this, various polyMOFs were generated from numerous MOF frameworks that were found by the team. The present work exemplifies how to employ the 1,4-benzene dicarboxylic acid units within the formed polymeric structures to attain stable polyMOF analogs analogous to the UiO-66 type. These polyUiO-66 structures were obtained from different organic polymers like Poly-2,5-bis(but-3-en-1-yloxy) terephthalic acid, Poly-2,5-bis(pent-4-en-1-yloxy) terephthalic acid, Poly-dimethyl 2,5-bis(hex-5-en-1-yloxy) terephthalate, and Dimethyl 2,5-bis(pent-4-en-1-yloxy) terephthalate, etc. PolyMOFs represent a hopeful path for creating MOF composites, as polymeric ligands allow precise control over the configuration, assembly, molecular weight, and loading of MOFs. Current work also shows that it influences the particle size, shape, and colloidal stability of MOFs and could improve the processability of these particles. It is the crystallization condition for forming MOF that after syntheses of polymeric ligands has to be established. In contrast to the traditional MOFs where the rigid small ligands give rise to well-ordered crystalline structures, the larger polymeric building blocks may induce chain entanglement and structural irregularities. This would give scope for forming amorphous materials with special applications. The molecular weight and backbone characteristics of the polymeric ligand can further influence the possibility of attaining the crystalline MOF structure. For example, crystalline stages can be accessed with alkyl linkers from three to eight carbons, such as in polyMOFs. Yet, this approach has not broadly been applied to other MOF frameworks and compounds through longer alkyl manacles or else with added efficient assemblies of interest (Ayala et al., 2017).

2.3 Post-synthetic Modification of Polymers on MOFs

Polymer chains might be attached on top of MOF frameworks. For instance, due to its large size, PNIPAM was grafted onto UiO-66-NH_2 through a reaction through the amine groups on the 2-amino-1,4-benzenedicarboxylate ligand, resulting in a thermoresponsive compound, NH_2-UiO-66@PNIPAM. This composite allows for controlled release by regulating the opening and closing of MOF pores based on temperature. Below the PNIPAM's transition temperature, the MOF remains accessible to guest molecules. However, above this temperature, PNIPAM coils block the pores and restrict the transport of guests. This research has spurred further investigation into MOF composites doped with thermoresponsive polymers, leading to the incorporation of novel functional properties. Liu and coworkers prepared NH_2-MIL101(Al) and attached polymers whose hydrophilic-to-hydrophobic transition at improved temperatures afford greater switch in the scattering or accumulation state of the MOF crystal in solution. They employed water-soluble polymers, for example, 2-(2-methoxyethoxy) ethyl methacrylate and oligo(ethylene glycol) methacrylate to prepare MIL101(Al)@poly(MEO_2MA-co-OEGMA) in their work (Liu et al., 2015). Li et al. designed a complex, multi-step process for grafting polymer chains onto MOFs using a combination of techniques called PSM, ATRP, and click chemistry. The technique developed in the paper differs from other straightforward grafting techniques. The authors initiated the process by functionalizing UiO-66-NH_2 with alkynyl groups followed by a click reaction to attach azide-terminated poly (tertabutyl methacrylate) PtBMA to the surface of the MOF. This produced UiO-66-NH_2@PtBMA, which was subsequently treated with an additional trifluoroacetic acid to produce poly (methacrylic acid) PMAA (Li et al., 2018).

2.4 Post-synthetic Conversion of Polymeric Organic Ligands

The fourth alternative route is the use of pre-formed polymers within already synthesized MOFs by simply incorporating them. Composites can be prepared in cases where direct MOF formation with polymeric ligands is infeasible or when the MOF is mismatched with the situations necessary for polymerization within or outside its pore space. Uemura et al. first described polymer melt processing using PEG (Kitao et al., 2017; Le Ouay et al., 2018). This pioneering work catalyzed massive studies on the introduction of polymers via melt or liquid routes, which is prompted by simulations indicating the spontaneous propagation of polymer chains in the pores of MOFs. The melt- or solvent-assisted polymer inclusion method uses a melting mediator. In 2016, Jones et al. incorporated polyethyleneimine (PEI) into MIL-101(Cr) to create MIL-101(Cr)@PEI, which demonstrated highly efficient direct air CO_2 capture due to PEI's high amine density, low volatility, and commercial availability (Darunte et al., 2016). The polymer was introduced to the MOF through a refluxing process in a toluene solution containing pre-formed PEI. Matzger et al. also used a similar approach wherein MOFs are soaked in a polymer-solvent solution to allow polymer introduction. While this technique remained employed to introduce biopolymers hooked on MOFs, it presents challenges as the extent of infusion is influenced by complex connections involving the polymer, the MOF, and the solvent, chlorobenzene (Koh et al., 2010).

2.5 Self-Assembly of MOFs in Conjunction with Polymers

Crystallizing MOFs around polymers is the other special attention-gaining method. Zhu et al. presented a novel method that utilizes metal ions (Zn^{2+}, Co^{2+}, Cu^{2+}, and Fe^{3+}) to crosslink alginate hydrogels. Metal ion-doped hydrogel fibers were suspended in ligand solutions to form polymer-MOFs. Metallic hydrogels were used to synthesize these MOFs by dipping them in different ligand solutions. The cross-linking density among MOF and alginate polymer chains would influence the hydrolytic constancy and surface area of the MOFs. Lee et al. demonstrated that MOFs in polymer-MOF beads, which are crystallized by soaking Cu^{2+}-cross-linked alginic acid hydrogels in ligand solutions, are hydrolytically stable and usable for seawater desalination (Lee et al., 2020; Zhu et al., 2016). In Section 2.2 (Lee et al., 2020), researchers focus on the direct application of pre-made polymeric ligands in the combination of MOFs, ensuring that polymers are combined into the framework during the crystallization process. In Section 2.3, describes the preparation of Polymer/MOF composites, where the initial step involves the synthesis of MOFs utilizing ligands that possess reactive functionalities such as C=C bonds, N_3, $-NH_2$, –OH, Cl, Br, and –CHO. Additionally, the components integrated into the frameworks of the MOF unit are achieved through polymerization utilizing established organic methodologies. Furthermore, Section 2.4 explains the second route and gives a general method of how it was possible, by direct deposition of prepared, pre-existing polymers or polymeric molecules into predetermined, prepared MOFs at the stage following synthesis but before completion and functional integration. The “ship in a bottle” effect, wherein polymers can be affixed to the inner surface of the MOF or encapsulated, elucidates those certain challenges, including the uneven distribution of the polymer and the obstruction of the MOF’s pores, are effectively tackled.

3. Composite of Polymer and Metal-Organic Frameworks

3.1 PolyMOF Composites

The hybridization of metal-organic frameworks (MOFs) and polymers represents a significant area of research, resulting in the creation of advanced materials including consumable MOF compounds, polymer-grafted MOFs, MOFs with polymer-filled pores, and structured MOFs. These composite materials demonstrate distinctive crystal-like, three-dimensional structures, differentiating them from conventional MOFs. PolyMOFs are a mixture of materials consisting of polymeric ligands and multitopic metal-coordinating monomers. A new strategy has been introduced for creating porous, crystalline polyMOFs by using custom-designed poly(pentenamer)s that carry H_2bdc units as pendant groups. Through ring-opening metathesis polymerization (ROMP), polymer ligands can be synthesized with controlled composition and dispersity (Đ). These poly(pentenamer)s, when combined with Zn(II), form polyMOFs that resemble IRMOF structures. Increasing the content of

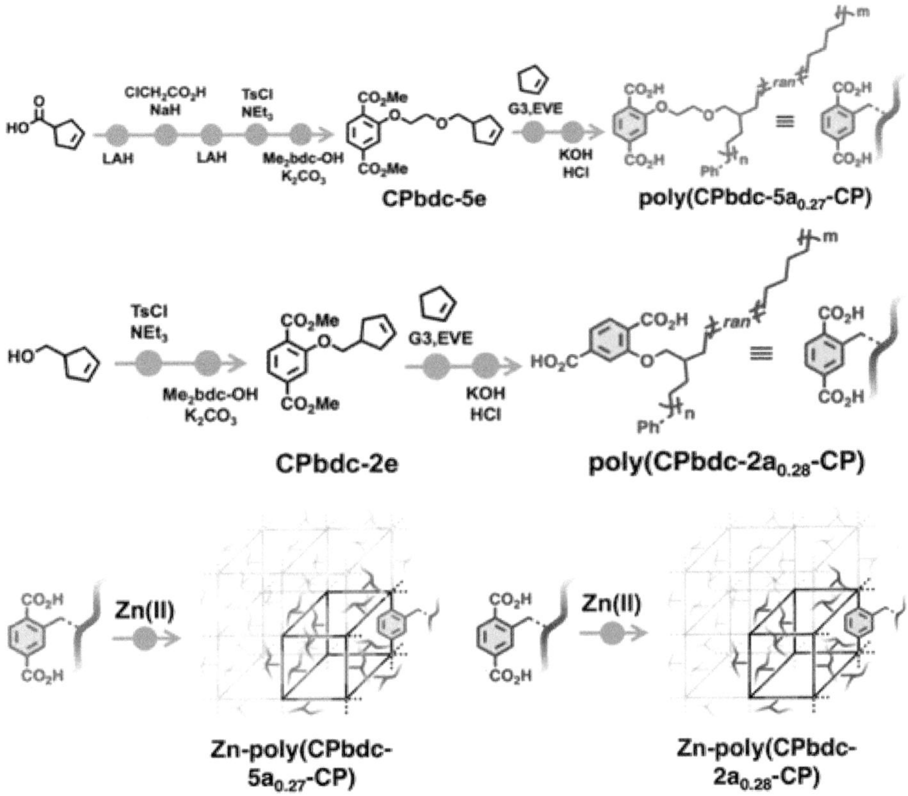

Fig. 3. Upper row: structure for the synthesis of poly (CPbdc-$5a_{0.27}$-CP). Middle row: the combination of poly (CPbdc-$2a_{0.28}$-CP). Lowest row: mixture of Zn-poly (CPbdc-$5a_{0.27}$-CP) (bottom-left) and Zn-poly(CPbdc-$2a_{0.28}$-CP) (bottom-right) under solvothermal conditions from poly(CPbdc-$5a_{0.27}$-CP) and poly(CPbdc-$2a_{0.28}$-CP). Adapted with authorization from (Hosaka et al., 2020a), 2020, American Chemical Society.

H_2bdc pendant ligands within the poly(pentenamer) chain enhances the crystallinity of the polyMOFs, which can be adjusted by varying the (co)monomer feed ratio during polymerization. ROMP further enables the creation of random terpolymers with controlled amounts of hydroxymethyl units alongside H2bdc pendants, which assemble efficiently into IRMOF-like polyMOFs. PolyMOFs formed from poly(pentenamer) linkers with shorter spacer lengths show distinct morphologies and exhibit significantly greater porosity compared to those with longer-spaced side-chain ligands. With ROMP's broad applicability, the range of available alkenamers, and the potential for side-chain modification, this approach represents a promising advancement for developing crystalline, porous polyMOFs with customized properties (Fig. 3) (Hosaka et al., 2020a).

3.2 Polymers in MOFs

Organic polymers possess characteristics such as ease of preparation, low weight, and good thermal and chemical stability. These are, therefore, ideal candidates to be combined with additional functional materials for the production of advanced composites. One feature of polymers confined in a nanometer-scale environment is the display of some intriguing properties that differ significantly from their bulk counterparts. A new approach has, therefore, emerged to exactly control polymer synthesis within porous MOFs. By integrating various MOFs and organic polymers, a new-class of material with synergistic properties was obtained to produce composites of MOF and organic polymer. These materials include polymer-MOF composites, which exhibit favourable fine-tuning of material structure by the inclusion of polymer. Mixed matrix membranes comprise MOF phases incorporated into a continuous polymeric matrix, and they are preferable to traditional adsorbents since they can transport higher levels of gas. Low-cost compound electrocatalysts of conductive polymers and MOFs exhibit the versatility of such hybrid materials. In recent years, research has been carried out to search for various configurations of MOF/polymer composites. Here, in the case of MOF/Polymer complexes, the polymer is located external to the structure of the MOF. On the other hand, in Polymer/MOF complexes, the polymer gets trapped exclusively in the pores of the MOF. The two architectures, therefore, offer opportunities for tuning material properties according to application needs (Lin et al., 2020).

3.2.1 MOF Composites in Polymers

Uniform porosity, tunable channel diameters, and customizable surface functionalities are characteristics that make MOFs promising platforms for controlled synthesis. Nano-sized pores act as programmable nano-reactors to precisely assemble polymers and MOFs with tailored properties. Integration of polymers and MOFs is an emerging area that spans the realm of polymer and MOF chemistry. In organic polymer@ MOF composites, the absorbent MOFs have unique pore characteristics: size, shape, dimension, and surface functionality can be exploited as confined reactors for polymerization. With the functionalization of active sites on pore surfaces, they may become templates and catalysts for polymer synthesis. Control over the polymerization of monomers encapsulated in nanochannels of MOF, such as styrene, methyl methacrylate, acrylonitrile, and pyrrole/thiophene derivatives, has allowed the formation of polymers with interesting prime and higher-order assemblies. The pore distance, geometry of the channels, and surface situation of the MOF host significantly distress chain alignment and elongation reaction kinetics of the monomers so that novel polymers become accessible (Schmidt, 2020).

Fe-BTC/PpPDA is a groundbreaking redox-active MOF-polymer composite that efficiently extracts and concentrates trace Au^{3+} ions from complex solutions. It removes over 90–99% of trace gold from various sources, such as surface water, wastewater, e-waste, and sewage ash. With high porosity, Fe-BTC/PpPDA achieves equilibrium within just 2 minutes. Upon regeneration, the composite enables gold concentration with a purity of 23.9 karats, setting a new standard. This high purity results from a dual adsorption-reduction mechanism that prevents ions like Cu^{2+} and

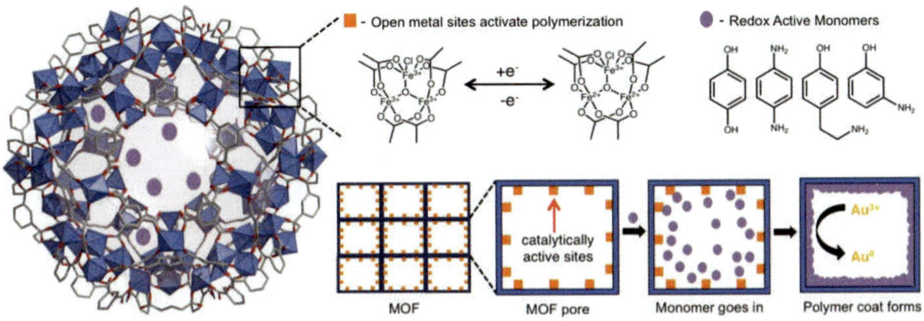

Fig. 4. A blueprint for designing innovative redox-active composites is shown. On the left is a polyhedral representation of the large cage structure within Fe-BTC. On the right, monomers diffuse into the porous network with accessible Fe^{3+} metal sites, which are reduced to Fe^{2+}, facilitating the polymerization process within the MOF's pores. Adapted with authorization from (Sun et al., 2018), 2018, American Chemical Society.

Ni^{2+} from interfering in the process. These properties are due to the added extrinsic porosity within the redox-active polymer. The study suggests that Fe-BTC/PpPDA could significantly impact gold recovery, and with selective MOF and polymer choices, this platform may be adapted to capture other high-value substances or contaminants from liquid or air (Fig. 4) (Sun et al., 2018).

3.2.2 Polymer/MOF Complexes

MOF/polymer composites, offering both the highly porous properties of MOFs and polymer flexibility, have recently opened exciting avenues to overcome weaknesses or improve the performance of MOFs in many different applications. The interaction of MOFs with polymers is highly diverse, depending upon weak host-guest connections, hydrogen bonding, π-π stacking, electrostatic interactions, organization between the polymer and the metal ion within the MOF structure, and covalent attachment of the polymer chains directly to the MOF ligands through "grafting to" or "grafting from" techniques. NMOFs, with tens to hundreds of nanometers in particle dimensions, possess great potential for a large number of applications. The surface alteration of NMOFs with organic polymers has numerous advantages, including improved performance and enhanced stability. This can be obtained by the development of an organic polymer shell to form core-shell nanostructures or through the dispersion of NMOF particles within macroporous polymer matrices configured as monoliths or beads. An effective way of enhancing biocompatibility and retarding the dissolution of the framework is the coating of NMOFs with an organic polymer coating. The polymer molecules for these methods should be capable of bonding to the NMOF particles either through the vacant metal coordination sites, electrostatic attraction, or covalent accessory to the linking ligands. Xu et al. have designed nanochannel-confined polypyrrole (PPy)/ZIF-8 nanocomposites exhibiting extremely tunable EC belongings (Xu et al., 2021).

4. Composites of Polymer MOFs and Metal-Ion Batteries

Although five distinct methods have been applied for the synthesis of polymer-MOF composites, their generic role in metal-ion batteries is the same. However, the selection of polymer greatly affects the conductivity and other properties of MOFs. Polymers are selected in such a way that the specific requirements for application are optimized. Apart from acting as structural templates, polymers can transform themselves into carbon materials that may act as conductive layers to shield unstable electrode materials. Metal-ion batteries are one of the primary types of energy-storing devices in which chemical energy is converted to electrical energy. The materials also have two subdivisions, namely, chemistry and construction. In this regard, sodium-ion batteries, for instance, will benefit from the synthesis of conductive polymers, such as polypyrrole, within the structure of MOFs (Liang, 2020a). This was shown by Dou et al. when the composite NMHFC@6PPy, prepared through the doping of conductive polypyrrole over ClO_4^- with $Na_{1+x}MnFe(CN)_6$, was used as SIB cathodes. Adding PPy significantly enhanced the electronic conductivity and the rate capability of NMHFC material. Another promising energy storing system is the potassium-ion battery (PIB), owing to the vast resources available, the low price, and lithium-like redox potential, along with the efficiency of potassium-ion transportation. Complex MOF-polymer complexes and their alteration into metal species or carbon derivatives have proven effective as potassium storage materials. Inspired by the promise of PIBs, the use of MOF/polyacrylonitrile (PAN) nanofiber results has been explored aimed at electrospinning applications (Hosaka et al., 2020b; Hu et al., 2017). Potassium-ion batteries inspired MOF/polyacrylonitrile (PAN) nanofiber derivatives for electrospinning. Zhang's team created electrospun MOF/PAN composites using ZIF-67 nanocubes, PAN, and DMF precursors, as illustrated in Fig. 5. Through carbonization and selenylation, $Co_{0.85}Se$@C nanoboxes were encapsulated within carbon nanofibers. These high-surface-area $Co_{0.85}Se$@C nanoboxes, with ample internal void space, help mitigate volume expansion, thereby enhancing cycling stability (Mondal et al., 2024).

Secondary batteries are mainly rechargeable metal-ion batteries that convert electrical energy into chemical energy, store it for some time, and then convert it back into electricity. In particular, LIBs represent the quintessential examples of secondary batteries that power everything from phones and laptops to electric cars and advanced energy storage devices. LIBs include four main elements such as an anode, a cathode, a separator, and an electrolyte. Charging and discharging are therefore very complex processes, involving the movement of electrons done in an external circuit, electrochemical responses within electrodes, and lithium-ion movement through the electrolyte. Upon charging, the cathode material passes its electrons and lithium particles to the circuit. Lithium ions are transported through the electrolyte to the anode. This electrochemical charge transmission between the cathode and anode enables the conversion and storage of electrical energy into chemical energy. However, in the case of discharge, electrode materials are subjected to electrochemical reactions to liberate the stored energy (Li et al., 2018).

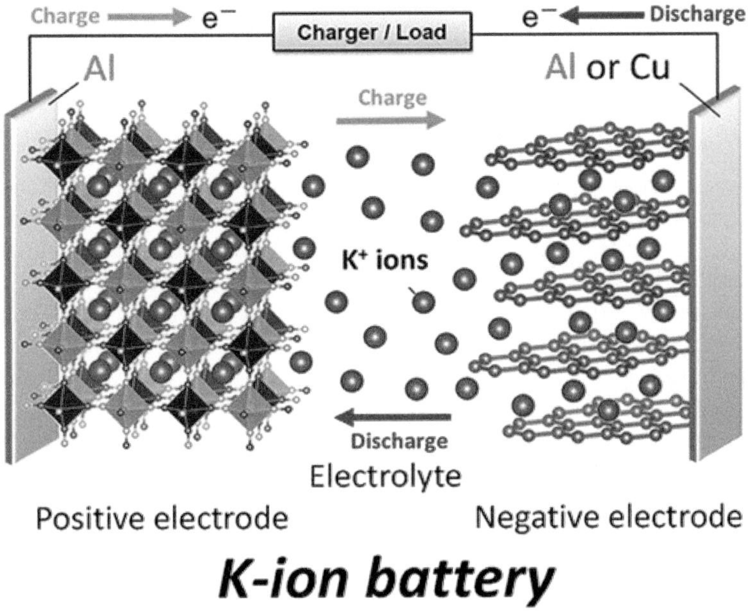

Fig. 5. Potassium Ion Batteries (PIBs). Adapted with authorization from (Mondal et al., 2024) 2020, American Chemical Society.

4.1 Composite of Polystyrene Sulfonate and MOFs

The Salt Lake brine's concentration of lithium, sodium, potassium, and magnesium ions has attracted several uses. They include battery making, ceramic materials, and light metal alloys. There, however, are critical challenges in the lithium ions extraction from these salt lake brines because ions of magnesium, sodium, and potassium are chemically similar to lithium ions and tend to interfere with this process. Polymer/MOF composites have become promising materials to overcome the challenge of selective lithium-ion extraction. They combine the processability of polymer membranes with the tunable pore structures and efficient surfaces of MOF membranes, hence making them more efficient and selective. Chen et al. developed a PSS@HKUST-1 membrane utilizing the abundant sulfonate clusters of polystyrene sulfonates (PSS) to identify alkaline and alkaline earth metal particles selectively. The composite membrane demonstrates outstanding lithium-ion conductivity, surpassing that of the pristine HKUST-1 membrane by five orders of magnitude. In addition, it possesses high selectivity over lithium ions against sodium, potassium, and magnesium ions with selectivity factors of 78, 99, and 10,296, respectively. The selectivity of the PSS@HKUST-1 membrane can be thought to be enhanced by size-sieving effects combined with specific interatomic affinities between the metal ions and the sulfonate groups. The pore size in HKUST-1 is about 0.9 nm, which is much bigger than the hydrated sizes of the lithium, sodium, and potassium ions; therefore, the size-sieving effect alone may not explain entirely the observed selectivity. Its sulfonate group holds anion ions with an order of binding affinity as such: $Li^+ < Na^+ < K^+$. Besides that, this PSS@HKUST-1 membrane polyethylene oxide blended shows much-improved water resistance than its counterpart in the pristine HKUST-1 membrane (Guo et al., 2016).

4.2 MOF/Polyethylene Oxide Composite

LIBs power a wide array of portable electronic devices, but there is a greater demand for LIBs as electric vehicles convert more mutually, especially with improved demands for higher power density and elevated safety standards. SSBs, with metallic lithium anodes and SSEs, promise to meet these needs. It would therefore be with theoretical size and small redox potential, promising it for significantly high energy densities, metallic lithium anodes. SSEs have improved safety features compared to traditional liquid electrolytes in that their flammability risks and risks of leakage can be circumvented, which reduces catastrophic short circuits. The physicochemical properties of SSEs play an important role in defining the overall electrochemical performance of SSBs. In this technology, SSEs are considered the main constituent; hence, high-performance SSE design is necessary to completely harness this technology. Very recently, Zhao and coworkers have proposed a new concept where a cobalt-based MOF, named ZIF-67, is used as a practical filler in complex polymer electrolytes in SSBs. They exhibit efficient transport of lithium ions by doping ZIF-67 into CPEs to open up possibilities for designing advanced SSBs (Blanc et al., 2020; Lu et al., 2022).

4.3 MOF/Polyphenol Formaldehyde Composite

Sung et al. detailed a clear method for the direct creation of M/MO/MC@C composites from MOFs: HKUST-1, MOF-74, and ZIF-8 via vapor phase polymerization (VPP) of phenol-formaldehyde (PF) on accessible metal sites. The MOFs have preserved their original morphology and crystallinity in PF@MOF composites, which exhibit controlled porosity and content. Under appropriate thermal treatment, thermolysis of such PF@MOF composites provided the respective M/MO/MC@C composites having varied metal phases of metals, metal oxides, or metal carbide phases along with crystallinity of the different ranges. The final phase and the crystallinity of the composite were determined by the specific metal ion and the thermolysis temperature. For instance, the Cu^{2+} ions were reduced to metallic copper (Cu@C), while the Mg^{2+} and Zn^{2+} ions were transformed to metal oxide nanoparticles (MgO@C and ZnO@C). Iron and cobalt ions were more complex as they provided metal oxides at relatively lower temperatures, such as Fe_2O_3 and CoO, whereas metal and metal carbide phases like Fe, Fe_3C, and Co were provided at higher temperatures. The data are evaluated as anode materials for LIBs in detail. M/MO/MC@C possessed great electrochemical performance and a very long cycling life as well as a high specific capacity. Especially, an outstanding specific capacity of 410 mAh g^{-1} at 1000 cycles of 1000 mA g^{-1} could be obtained from CuO@C This is more than two-fold the capacity compared to that obtained from the relatively less-optimized sample with the same material directly oxidized from CuO. Data shows applying MOF-derived composites for the fabrication of high-performance LIB anode materials has good potential (Kang et al., 2018).

5. Conclusions

The need for innovative solutions has been driven by the limitations of bits of knowledge related to current battery systems. Novel materials with unique structural and chemical properties are being developed to address this need. Hybrid materials combining organic polymers and metal-organic frameworks (MOFs) are a promising avenue. Materials must have certain key characteristics: high porosity, excellent electrical and ionic conduction, robust physical and chemical stability, and scalability. Energy storage batteries must, at optimum performance, focus primarily on physicochemical stability as the active material may crack under volume changes during performance. Efficient electron and ion transport is also paramount because internal resistance and related loss in voltage should not increase. High ionic conductivity is particularly important when seeking to maximize charge storage. Polymer-MOF composites can be synthesized via different methods. *In situ* polymerization, formation of polyMOF, post-synthetic modification, and self-assembly lead to a variety of preparation methods of different composites. Configurations of PolyMOF, Polymer@MOF, and MOF@polymer. MOFs and hybrid polymer-based materials are expected to be very promising systems for metal-ion batteries, as they have lower flammability, significant mechanical rigidity, and a wide electrochemical window, thereby contributing to safe operation and high energy density, which eliminates risks of leakage, short circuits, or thermal runaway. While there is much progress, there is still much to be done in the form of research to report tasks on polymer–MOF composites, recognizing notable advancements. Unlocking the full potential of these materials for advanced energy storage applications can be achieved by exploring novel materials and innovative synthesis techniques.

References

Abbasi, N., Moradi, M., Hajati, S., Kiani, M.A. and Toth, J. 2017. *In-situ* growth of ultrathin Ni6MnO8 nanosheets on nickel foam as a binder-free positive electrode for asymmetric supercapacitor: Effects of alkaline aqueous and redox additive electrolytes. J. Mol. Liq. 244: 269–278. https://doi.org/10.1016/j.molliq.2017.09.018.

Ayala, S., Zhang, Z. and Cohen, S.M. 2017. Hierarchical structure and porosity in UiO-66 polyMOFs. Chemical Communications 53: 3058–3061. https://doi.org/10.1039/C6CC10225E.

Blanc, L.E., Kundu, D. and Nazar, L.F. 2020. Scientific challenges for the implementation of Zn-Ion batteries. Joule 4: 771–799. https://doi.org/10.1016/j.joule.2020.03.002.

Chameh, B., Moradi, M. and Kaveian, S. 2020. Synthesis of hybrid ZIF-derived binary ZnS/CoS composite as high areal-capacitance supercapacitor. Synth. Met. 260: 116262. https://doi.org/10.1016/j.synthmet.2019.116262.

Chameh, B., Moradi, M., Hajati, S. and Hessari, F.A. 2021. Design and construction of ZIF(8 and 67) supported Fe_3O_4 composite as advanced materials of high performance supercapacitor. Physica E Low Dimens Syst. Nanostruct. 126: 114442. https://doi.org/10.1016/j.physe.2020.114442.

Costa, C.M., Lizundia, E. and Lanceros-Méndez, S. 2020. Polymers for advanced lithium-ion batteries: State of the art and future needs on polymers for the different battery components. Prog. Energy Combust. Sci. 79: 100846. https://doi.org/10.1016/j.pecs.2020.100846.

Darunte, L.A., Oetomo, A.D., Walton, K.S., Sholl, D.S. and Jones, C.W. 2016. Direct air capture of CO_2 using amine functionalized MIL-101(Cr). ACS Sustain Chem. Eng. 4: 5761–5768. https://doi.org/10.1021/acssuschemeng.6b01692.

Ding, N., Li, H., Feng, X., Wang, Q., Wang, S., Ma, L. et al. 2016. Partitioning MOF-5 into confined and hydrophobic compartments for carbon capture under humid conditions. J. Am. Chem. Soc. 138: 10100–10103. https://doi.org/10.1021/jacs.6b06051.

Gamage, N.-D.H., McDonald, K.A. and Matzger, A.J. 2016. MOF-5-polystyrene: direct production from monomer, improved hydrolytic stability, and unique guest adsorption. Angewandte Chemie International Edition 55: 12099–12103. https://doi.org/10.1002/anie.201606926.

Guo, Y., Ying, Y., Mao, Y., Peng, X. and Chen, B. 2016. Polystyrene sulfonate threaded through a metal–organic framework membrane for fast and selective lithium-ion separation. Angewandte Chemie International Edition 55: 15120–15124. https://doi.org/10.1002/anie.201607329.

Hosaka, T., Kubota, K., Hameed, A.S. and Komaba, S. 2020a. Research development on K-ion batteries. Chem. Rev. 120: 6358–6466. https://doi.org/10.1021/acs.chemrev.9b00463.

Hu, F., Li, L. and Jiang, X. 2017. Hierarchical octahedral NaMnFe(CN)6 and NaMnFe(CN)6@Ppy as cathode materials for sodium-ion batteries. Chin. J. Chem. 35: 415–419. https://doi.org/10.1002/cjoc.201600713.

Kang, M.S., Lee, D.-H., Lee, K.-J., Kim, H.S., Ahn, J., Sung, Y.-E. et al. 2018. Porosity- and content-controlled metal/metal oxide/metal carbide@carbon (M/MO/MC@C) composites derived from MOFs: mechanism study and application for lithium-ion batteries. New Journal of Chemistry 42: 18678–18689. https://doi.org/10.1039/C8NJ04919J.

Kitao, T., Zhang, Y., Kitagawa, S., Wang, B. and Uemura, T. 2017. Hybridization of MOFs and polymers. Chem. Soc. Rev. 46: 3108–3133. https://doi.org/10.1039/C7CS00041C.

Koh, K., Wong-Foy, A.G. and Matzger, A.J. 2010. Coordination copolymerization mediated by Zn4O(CO2R)6 metal clusters: a balancing act between statistics and geometry. J. Am. Chem. Soc. 132: 15005–15010. https://doi.org/10.1021/ja1065009.

Krishan, O. and Suhag, S. 2019. An updated review of energy storage systems: Classification and applications in distributed generation power systems incorporating renewable energy resources. Int. J. Energy Res. 43: 6171–6210. https://doi.org/10.1002/er.4285.

Le Ouay, B., Watanabe, C., Mochizuki, S., Takayanagi, M., Nagaoka, M., Kitao, T. et al. 2018. Selective sorting of polymers with different terminal groups using metal-organic frameworks. Nat. Commun. 9: 3635. https://doi.org/10.1038/s41467-018-06099-z.

Lee, S.J., Hann, T. and Park, S.H. 2020. Seawater desalination using MOF-incorporated Cu-based alginate beads without energy consumption. ACS Appl. Mater. Interfaces 12: 16319–16326. https://doi.org/10.1021/acsami.9b22843.

Li, M., Lu, J., Chen, Z. and Amine, K. 2018. 30 Years of lithium-ion batteries. Advanced Materials 30: 1800561. https://doi.org/10.1002/adma.201800561.

Li, Y., Liu, J., Zhang, K., Lei, L. and Lei, Z. 2018. UiO-66-NH_2@PMAA: A hybrid polymer–MOFs architecture for pectinase immobilization. Ind. Eng. Chem. Res. 57: 559–567. https://doi.org/10.1021/acs.iecr.7b03398.

Liang, Y., Dong, H., Aurbach, D. and Yao, Y. 2020a. Current status and future directions of multivalent metal-ion batteries. Nat. Energy 5: 646–656. 10.1038/s41560-020-0655-0.

Lin, R., Hou, J., Li, M., Wang, Z., Ge, L., Li, S. et al. 2020. Interfacial engineering of a polymer–MOF composite by *in situ* vitrification. Chemical Communications 56: 3609–3612. https://doi.org/10.1039/D0CC00664E.

Liu, H., Zhu, H. and Zhu, S. 2015. Reversibly dispersible/collectable metal-organic frameworks prepared by grafting thermally responsive and switchable polymers. Macromol. Mater Eng. 300: 191–197. https://doi.org/10.1002/mame.201400251.

Lu, G., Wei, H., Shen, C., Zhou, F., Zhang, M., Chen, Y. et al. 2022. Bifunctional MOF doped PEO composite electrolyte for long-life cycle solid lithium ion battery. ACS Appl. Mater. Interfaces 14: 45476–45483. https://doi.org/10.1021/acsami.2c13613.

Mehtab, T., Yasin, G., Arif, M., Shakeel, M., Korai, R.M., Nadeem, M. et al. 2019. Metal-organic frameworks for energy storage devices: Batteries and supercapacitors. J. Energy Storage 21: 632–646. https://doi.org/10.1016/j.est.2018.12.025.

Mondal, P., Sensharma, D. and Cohen, S.M. 2024. Polymer–Metal–Organic Frameworks (polyMOFs) based on tailor-made Poly(alkenamer)s. Chemistry of Materials 36: 9696–9703. https://doi.org/10.1021/acs.chemmater.4c01794.

Pan, Z., Liu, X., Yang, J., Li, X., Liu, Z., Loh, X.J. et al. 2021. Aqueous rechargeable multivalent metal-ion batteries: advances and challenges. Adv. Energy Mater 11: 2100608. https://doi.org/10.1002/aenm.202100608.

Schmidt, B.V.K.J. 2020. Metal-organic frameworks in polymer science: polymerization catalysis, polymerization environment, and hybrid materials. Macromol. Rapid Commun. 41: 1900333. https://doi.org/10.1002/marc.201900333.

Sun, D.T., Gasilova, N., Yang, S., Oveisi, E. and Queen, W.L. 2018. Rapid, selective extraction of trace amounts of gold from complex water mixtures with a metal–organic framework (MOF)/polymer composite. J. Am. Chem. Soc. 140: 16697–16703. https://doi.org/10.1021/jacs.8b09555.

Uemura, T., Kitagawa, K., Horike, S., Kawamura, T., Kitagawa, S., Mizuno, M. et al. 2005. Radical polymerisation of styrene in porous coordination polymers. Chemical Communications 5968–5970. https://doi.org/10.1039/B508588H.

Xie, L.S., Skorupskii, G. and Dincă, M. 2020. Electrically conductive metal–organic frameworks. Chem. Rev. 120: 8536–8580. https://doi.org/10.1021/acs.chemrev.9b00766.

Xu, M.-M., Xie, L.-H. and Li, J.-R. 2021. Metal–organic framework/polymer hybrid materials. *In*: Wang, B. (ed.). Hybrid Metal–Organic Framework and Covalent Organic Framework Polymers. The Royal Society of Chemistry, p. 0. https://doi.org/10.1039/9781839163456-00072.

Yang, S., Karve, V.V., Justin, A., Kochetygov, I., Espín, J., Asgari, M. et al. 2021a. Enhancing MOF performance through the introduction of polymer guests. Coord. Chem. Rev. 427: 213525. https://doi.org/10.1016/j.ccr.2020.213525.

Zhang, Y., Feng, X., Yuan, S., Zhou, J. and Wang, B. 2016. Challenges and recent advances in MOF–polymer composite membranes for gas separation. Inorg. Chem. Front. 3: 896–909. https://doi.org/10.1039/C6QI00042H.

Zhang, Z., Nguyen, H.T.H., Miller, S.A. and Cohen, S.M. 2015. polyMOFs: A class of interconvertible polymer-metal-organic-framework hybrid materials. Angewandte Chemie International Edition 54: 6152–6157. https://doi.org/10.1002/anie.201502733.

Zhu, H., Zhang, Q. and Zhu, S. 2016. Alginate hydrogel: a shapeable and versatile platform for *in situ* preparation of metal–organic framework–polymer composites. ACS Appl. Mater. Interfaces 8: 17395–17401. https://doi.org/10.1021/acsami.6b04505.

Chapter 4

Anthraquinone-Based Redox Polymers for Organic Batteries

Monika Kumawat

1. Introduction

Organic batteries will outperform metal-based ones to meet global energy storage needs. Electric mobility favors lithium-ion batteries over transition metals (Kim et al., 2023). Advantages include flexible and lightweight designs, little environmental impact, and abundant and sustainable materials. Research is in progress to improve organic battery performance and anthraquinone-based redox polymer electrochemical properties. Low electrical conductivity in conjugated microporous polymers (CMP) frameworks affects charge transfer and electrochemical efficiency. Redox-active molecules like anthraquinone molecules increase energy density and theoretical capacity (Mohamed et al., 2024).

Reversible redox reactions charge and discharge organic batteries. Instead of finite, environmentally harmful metals, organic lithium-ion batteries use carbon molecules. Redox-active polymers could replace inorganic lithium-ion battery electrodes (Goujon et al., 2021). Sustainable manufacturing and disposal are possible. Energy storage technologies become sustainable and resource-preserving with the use of organic materials. A stable redox state allows polycyclic aromatic molecule anthraquinone (AQ) to store energy efficiently. Wide-range energy storage with aqueous organic redox flow batteries may be cheaper and safer than vanadium flow and lithium-ion systems (Wu et al., 2020). The quinone structure of AQ allows fast charge/discharge cycles and reversible redox reactions. The highest theoretical capacity of anthraquinone could boost battery energy density. This

Department of Chemistry, Faculty of Science, Bhupal Nobles' University, Udaipur, Rajasthan, India.
Email: monika.kumawat84@gmail.com

DOI: 10.1201/9781003546672-4

performance metric is crucial for electric cars and portable electronics. Researchers developed anthraquinone-based redox polymers for their benefits. These polymers combine anthraquinones' redox activity with mechanical stability and functionality. Polymer AQ integration improves small-molecule anthraquinone cycle stability and electrolyte solubility. Battery efficiency and lifespan are improved by AQ-binding redox polymers. Molecular design enhances anthraquinone-based redox polymers. Researchers can improve redox-active site connectivity, conductivity, and electrolyte interaction by changing polymer structure. Batteries outperform older technology in capacity, cycle life, and rate.

2. Anthraquinone-Based Redox Polymers

2.1 Anthraquinone Chemistry

Anthraquinones are abundant and natural. They're in bacteria, fungi, plants, and animals (Diaz-Munoz et al., 2018). Anthraquinone, a polycyclic aromatic hydrocarbon with the formula $C_{14}H_8O_2$, supports multiple organic battery redox-active compounds (Fig. 1). Anthraquinone and its derivatives are a class of polycyclic aromatic compounds useful for various purposes (Crespo, 2021). Two benzene rings connected by a quinone molecule provide an adaptable redox reaction. Its design allows energy storage-related reversible reduction and oxidation. The diketone group in anthraquinone experiences a two-electron redox process that transforms it from quinone to hydroquinone. The ability to efficiently receive and donate electrons makes anthraquinone-based molecules suitable for rechargeable batteries. A new organic cathode which is potassium 2, 6-dihydroxyanthraquinone is produced and it has been fully characterized for K-ion batteries (Xie et al., 2024).

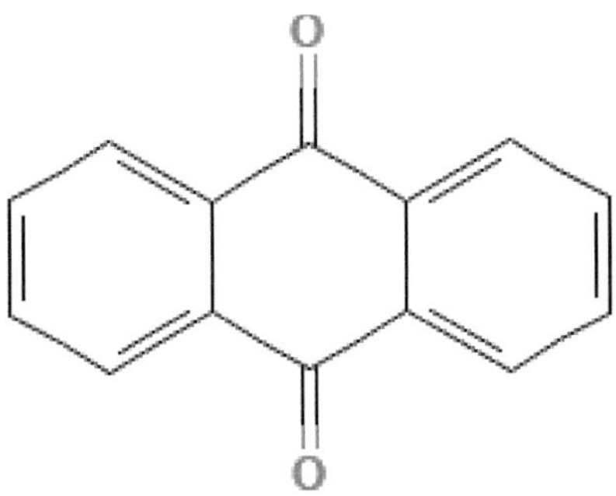

Fig. 1. Typical structure of anthraquinone.

2.2 Synthesis of Anthraquinone-Based Redox Polymers

Organically functioning anthraquinone (AQ) in polymer frameworks creates redox polymers. The key methods are Condensation Reactions, Coupling Reactions, Nucleophilic Aromatic Substitution Reactions, and the Diels-Alder Reaction. These synthesized ways provide unique ARP redox potential, solubility, and structural stability modification.

2.2.1 Condensation Reaction

Anthraquinone-based redox Polymer synthesis uses condensation to form C-C, C-N, or C-O bonds. Some reactions remove water, ammonia, and alcohol. To make derivatives, 1,4-anthraquinone is condensed with aldehydes or ketones. Anaquinone with formaldehyde produces 2-methylene-anthraquinone. Functional groups form polymerization sites (Fig. 2).

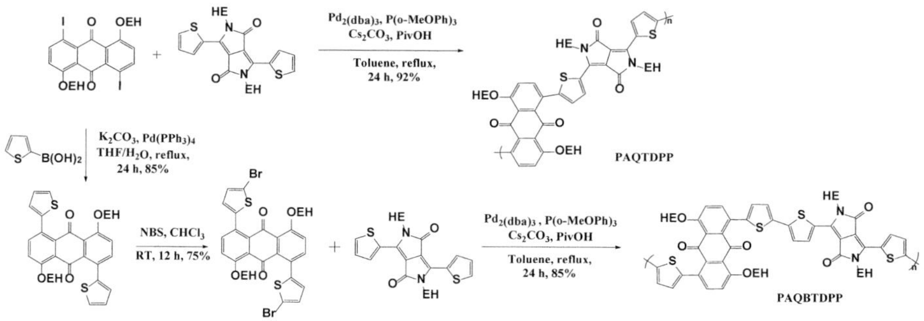

Fig. 2. Synthesis of anthraquinone-based organic polymers via condensation reaction.

2.2.2 Coupling Reactions

Covalent bonds between anthraquinone units and functional groups or monomers require coupling processes. Using substances and catalysts, reaction conditions can be optimized to achieve the desired polymer architecture. Some important couplings are Diazo-Coupling Polymerization, Sonogashira Coupling, Suzuki Miyaura Coupling, Buchwald-Hartwig Coupling, Oxidative Coupling (Fig. 3) Polymerization and Diazo-coupling polymerization.

2.2.3 Nucleophilic Aromatic Substitution Reactions

Nucleophilic aromatic substitution reactions (S_NAr) can replace anthraquinone halogens (Fig. 4). This method is appropriate for AQ-based polymers because electron donating or withdrawing components affect the polymer's redox potential.

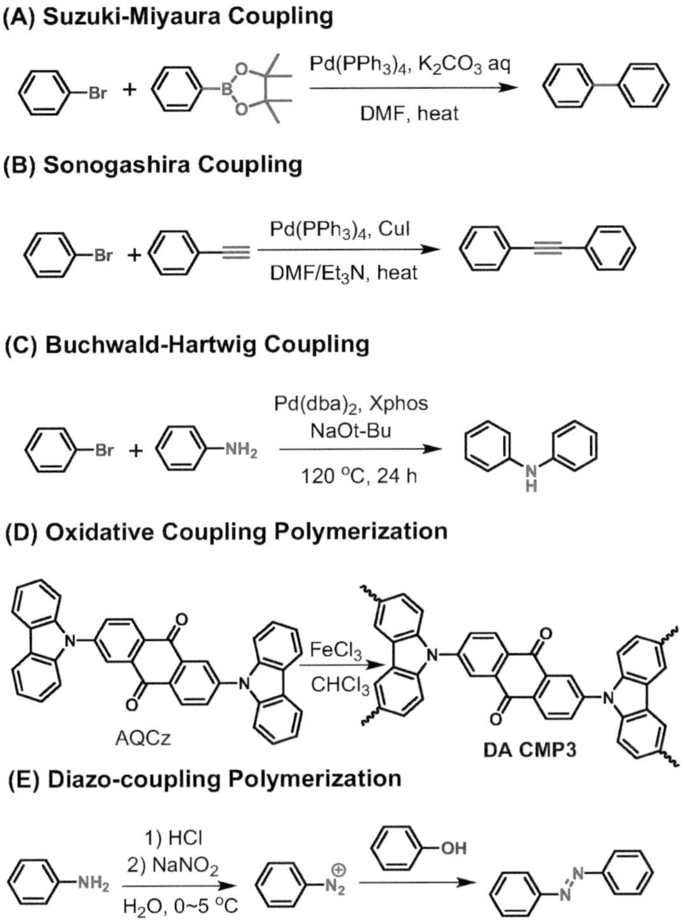

Fig. 3. Synthesis of anthraquinone-based porous organic polymers via various coupling reactions (Source: Qi et al., 2023).

2.2.4 Diels – Alder Reaction

Diels-Alder polymer AQ unit addition and fused-ring architectures work well. This [4+2] cycloaddition produces a cyclohexene ring from a diene and dienophile (Fig. 5). Dienophile anthroquinone interacts with dienes to form polycyclic AQ-based polymers.

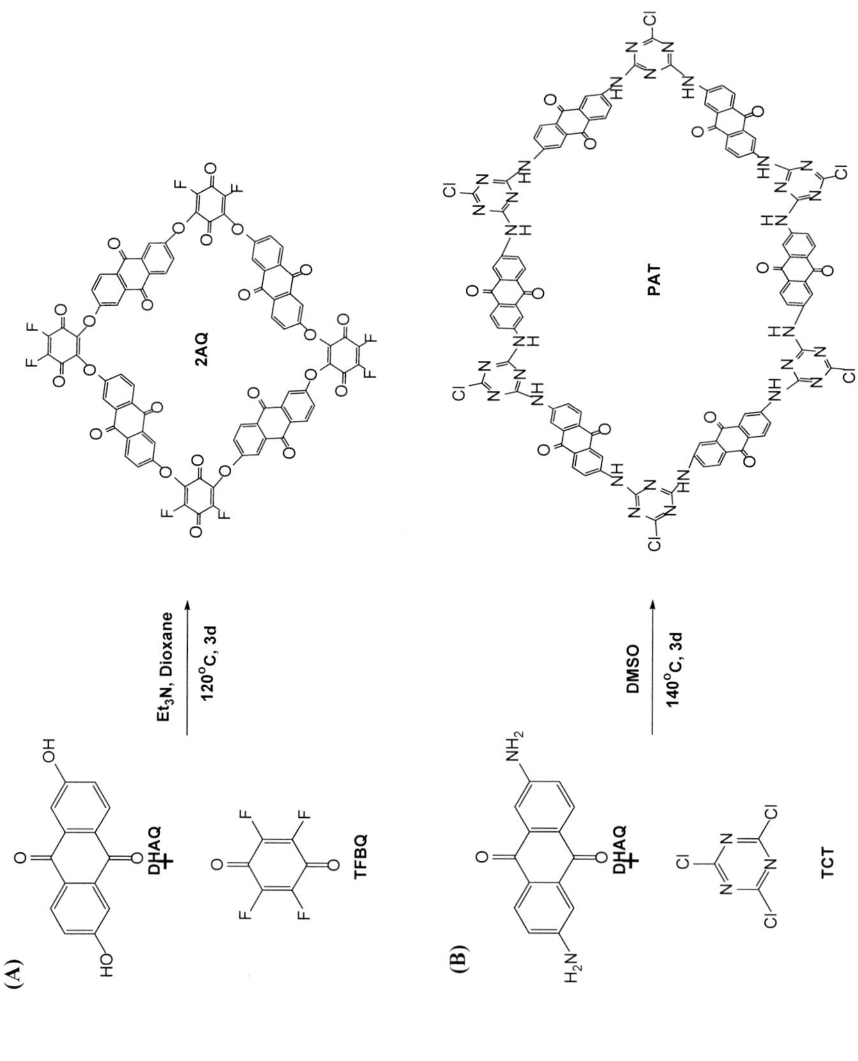

Fig. 4. Synthesis of anthraquinone-based porous organic polymers via nucleophilic aromatic substitution reaction.

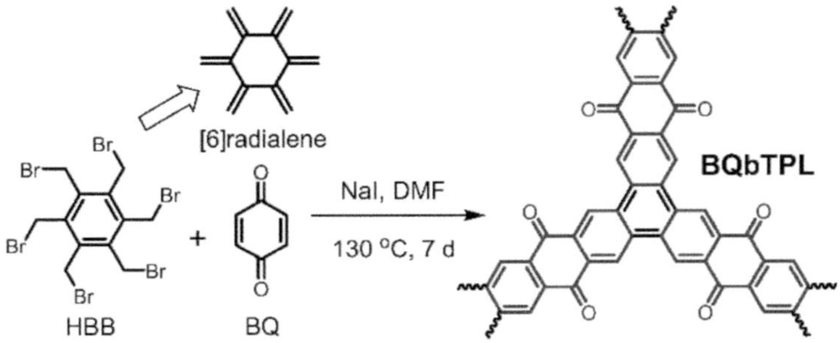

Fig. 5. Synthesis of ladder-type conjugated microporous polymer BQbTPL via Diels – Alder reaction (Source: Qi et al., 2023).

2.3 Structural Characterization

Molecular research, morphological studies, spectroscopic techniques, and crystallographic methods are all used together to figure out the structure of anthraquinone-based redox polymers. The electrochemical performance of AQ molecules is affected by crystal growth and the chemical surroundings, which includes adding a binder and conductive carbon. When the number of Electron Withdrawing Groups (EWGs) goes up, so does the reductional potential of anthraquinone. The general effect of compounds on the redox potentials of AQ is -C2F5 > -CF3 > -F > -Cl > -Br when the number of EWGs stays the same (Feng et al., 2021).

2.3.1 Fourier-Transform Infrared (FTIR) Spectroscopy

The functional groups in AQ-based polymers are identified by FTIR. Polymer structure contains AQ as seen by the AQ unit, carbonyl group, and polymer backbone absorption bands. FTIR spectrum deviations during cycling may indicate chemical degradation. A Fourier transform infrared absorption (FTIR) analyser measures impurities, surface molecular composition, structural and geometric isomers, polymer and solution orientation, and IR energy (Guerrero-Pérez and Patience, 2020).

2.3.2 Ultraviolet-visible (UV-Vis) Spectroscopy

UV-Vis spectroscopy tracks polymer AQ unit's electrical changes. Absorbance spectra show the polymer's electronic structure and conjugation length, which govern redox behaviour. The interaction between electromagnetic radiation (EMR) and matter at a certain wavelength is measured in this method (Akash and Rehman, 2020).

2.3.3 Nuclear Magnetic Resonance (NMR) Spectroscopy

NMR is necessary for AQ-based redox polymer molecular structure characterization. It reveals the polymer chain's chemical environment and AQ unit's structure and conformation. NMR is used to detect the chemical structure of the byproducts (Zhu et al., 2021). Proton (^{1}H) NMR and Carbon-13 (^{13}C) AQ groups and other functional moieties on polymer backbones are typically found using NMR. Integration of NMR spectra data, coupling constants, and chemical shifts determines polymerization, linkage type, and repeat unit structure.

2.3.4 Mass Spectrometry (MS)

MS, also known as Matrix-Assisted Laser Desorption/Ionization (MALDI) and Electrospray Ionisation (ESI) can confirm AQ units in polymers and provide molecular weight distributions. MS supports NMR by revealing molecular weight distribution, end groups, and contaminants or side products. MS plays an integral part in the development of techniques for the analysis and knowledge of complex, natural organic matter (NOM) type molecules (Letourneau and Volmer, 2023).

2.3.5 Gel Permeation Chromatography (GPC)

Size-Exclusion Chromatography (SEC), which is another name for GPC, is used to find the molecular weight and polydispersity index (PDI) of AQ-based polymers. This method helps us figure out how the lengths of polymer chains are spread out, which can change the electrochemical performance and physical features of the polymer. Most of the time, results are compared to bendable polymer standards, and this is the method most often used to measure molecular weight (Fair et al., 2021).

2.3.6 Scanning Electron Microscopy (SEM)

Polymer surface morphology is demonstrated by SEM. You can see polymer porosity, surface irregularity, and roughness. Due to its surface area, these morphological qualities explain the polymer's electrolyte interactions and redox processes. SEM is implemented to determine FRP void size, shape, volume fractions, and porosity (Rosemeier et al., 2022).

2.3.7 Transmission Electron Microscopy (TEM)

AQ-based redox polymers' interior structure is shown by nanoscale TEM. This method demonstrates AQ unit distribution and orientation in the polymer matrix, which impacts mechanical and electrochemical properties. The microstructure and composition can be examined via the use of energy-dispersive X-ray spectroscopy (EDX) combined with scanning transmission electron microscopy (STEM) (Lu and Elimelech, 2021).

2.3.8 Atomic Force Microscopy (AFM)

Nanometer-resolution polymer surface topography maps are made using AFM. Polymer battery behaviour must be understood since it evaluates surface roughness, phase separation, and mechanical properties like stiffness. Combining AFM with an infrared spectroscope (AFM-IR) improves chemical studies of polymers (Nguyen-Tri et al., 2020).

2.3.9 X-ray Diffraction (XRD)

XRD evaluates AQ-based redox polymers' crystals. It provides crystallinity, orientation, and structural characteristics. Crystalline domains change polymer stiffness and ion transport channels, impacting mechanical, thermodynamic, and electrochemical properties. XRD teaches us about the short-range order of non-crystalline materials and the long-range order of crystalline materials (Khan et al., 2020).

2.3.10 Single-Crystal X-ray Crystallography

Despite polymer limitations, this technique can provide atomic-level AQ unit positioning in the structure of crystals. It explains structural factors impacting polymer stability and redox reactivity. SCXRD is a fundamental chemical and materials science characterisation method with over one million organic and metal-organic structures (Schriber et al., 2022).

3. Electrochemical Properties

With their unique redox mechanisms, high energy density, beneficial voltage profiles, and stable cycle life, anthraquinone (AQ)-based redox polymers show promising electrochemical properties for application in organic batteries. These characteristics are necessary for the development of energy storage systems that are durable and efficient. High-performance supercapacitors benefit from conjugated microporous polymers' energy conversion, storage, and harnessing (Luo et al., 2021).

3.1 Redox Mechanisms

Redox reactions gain or lose electrons, changing oxidation state (Fig. 6) (Pan et al., 2023).

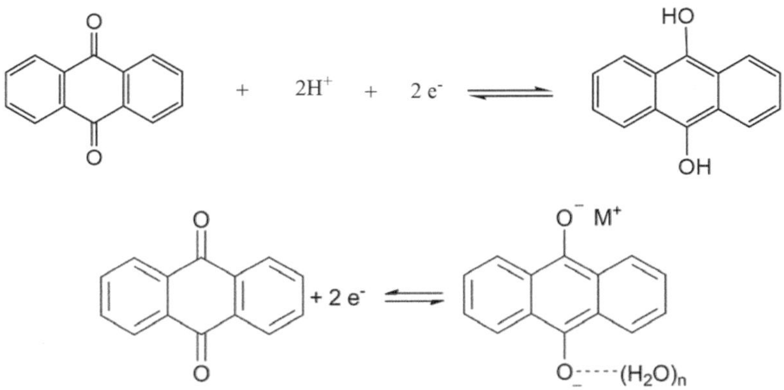

Fig. 6. Redox reactions of anthraquinone in acidic (top row) and alkaline (bottom row) electrolytes (Source: Thierry et al., 2018).

3.2 Quinone-Hydroquinone Redox Couple

Redox-active species in solution can be made more soluble by adding solubilizing chemicals (Fontmorin et al., 2022). Anthraquinone's principal redox process is the reversible conversion of quinone to hydroquinone. Quinone reduces to hydroquinone with protons after a two-electron transfer. A fuel cell based on the quinone-hydroquinone redox pair can advance the development of small organic molecule-based regenerative flow batteries (Nawar et al., 2013).

3.3 Conjugation and Electron Delocalization

Extended conjugation in anthraquinones effectively delocalizes electrons, stabilising redox states. Electron flow across the π-conjugated system enhances the reversibility and kinetics of redox reactions. Electronic delocalization enhances absorption and emission peak redshifts when compared to polymers with aromatic linkers (Yin et al., 2021). In substituted AQs, intra and intermolecular hydrogen bonding and electron flow complement each other. AQ structure-charge distribution and intramolecular HB effects indirectly affect hydrogen bonding (Nakayama et al., 2020). For high redox activity and capacity, AQ-based materials need electron delocalization. The electron delocalization of withdrawing compounds is accessible to boost intrinsic electronic conductivity (Su et al., 2024).

3.4 Functional Group Modifications

Functionalizing AQ with electron-donating or electron-withdrawing groups modifies its redox properties. Alkyl or alkoxy groups improve electron density on the AQ core, lowering the redox potential and increasing reducing power. Halogens and nitro groups decrease electron density and raise redox potential. Functional group change improves water absorption (Xie et al., 2024). Adjustments alter the redox potential for battery applications. The hydro-quinone functional group performed less catalytically than the anthraquinone functional groups of AQs on TDAG breakdown.

3.5 Energy Density and Voltage Profiles

The theoretical capacity (number of electrons transferred per molecule) and practical capacity (real-world redox process efficiency) influence AQ-based redox polymers' energy density. AQ's theoretical capacity utilizes a two-electron redox method. In a polymer matrix, polymer shape, electrode architecture, and electrolyte interactions affect capacity. Redox-active polymer architectures for active materials are abundant in organic materials due to their structural variety (Hager et al., 2020).

The voltage profile of AQ-based redox polymers depends on the AQ/AH pair's redox potential. For redox transitions, charge and discharge cycles produce voltage plateaus. Proper positioning and scale of distributed generation are needed to enhance the voltage profile and decrease electrical power losses (Fig. 7) (Ntombela et al., 2022). These plateaus provide stable voltage for energy generation. Organic materials undergoing unique electrochemical conversion processes can meet metal-ion (Li^+, Na^+, K^+, and Mg^{2+})/metal battery voltage, capacity, cycling stability, and rate performance (Wang et al., 2022).

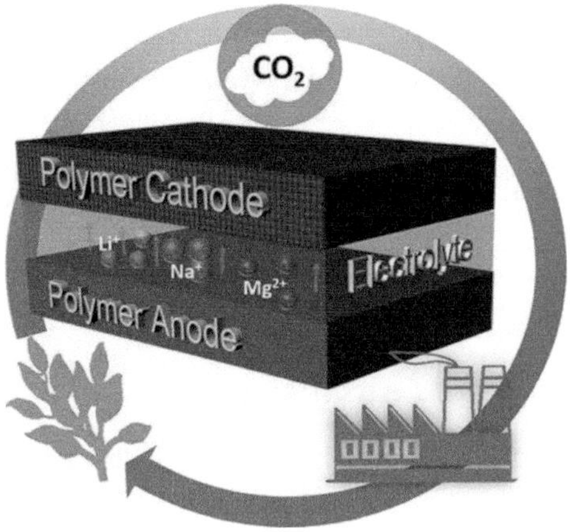

Fig. 7. Redox-active polymers (redoxmers) for electrochemical energy storage (Source: Yang et al., 2019).

3.6 Polymer and Composite Effects

Flexibility, lightweight, and programmable redox properties distinguish AQ-based redox polymers from inorganic materials. AQ-based polymers have a lower energy density than inorganic materials; their environmentally benign procedures and ability to change the AQ structure make them attractive for new energy storage devices (Fig. 8). Only a small percentage of the polymer matrix is required to provide stability and to satisfy various kinds of applications (Masek et al., 2021).

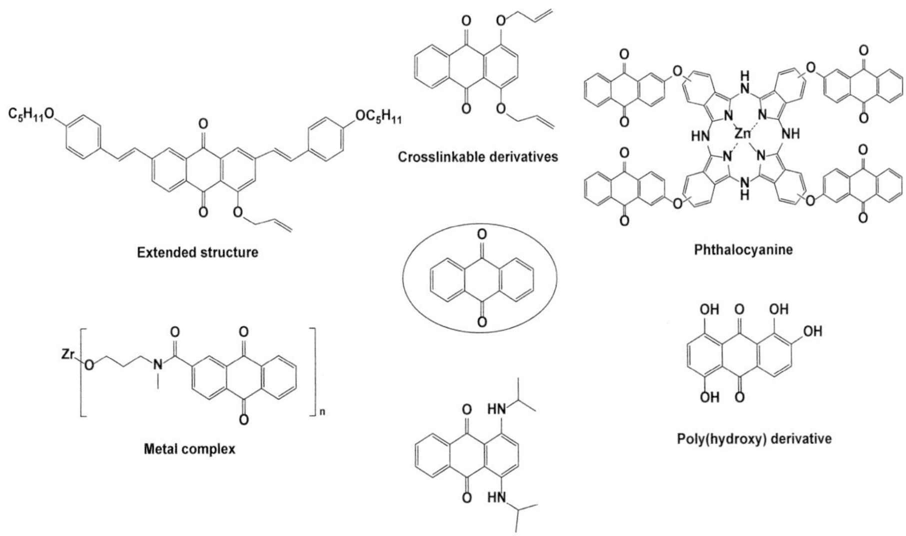

Fig. 8. Anthraquinones as photoinitiators of polymerization (Source: Dumur, 2023).

3.7 Electrochemical Stability

The efficiency of anthraquinone-based redox polymers in batteries is significantly influenced by their electrochemical stability. The robustness of the matrix and the reversibility of the AQ/AH redox couple are both factors that influence stability. Side reactions, including the formation of inactive species or degradation of the polymer backbone, can be a result of repeated cycling. Conjugated AQ is stable in degradation. Flexibility, safety, and longevity make redox flow batteries (RFBs) ideal for large-scale storage (Amini et al., 2023).

3.8 Polymer Backbone and Crosslinking

Choosing the right polymer backbone and crosslinking methods can improve AQ-based redox polymers' stability and cycle life. Strong connections are prevented by the rigid polymer backbone (Li et al., 2023). Stiff polymers or crosslinked networks preserve electrode material by preventing electrolyte expansion or dissolution. Redox reactions change the charge of active molecules, affecting solubility, so crosslinked polymers can help (Hager et al., 2020). Immobilising AQ units with crosslinking reduces mobility and harm. Porous structures improve electrolyte-redox site interaction. Highly crosslinked network structures and rigid polymer backbones make POPs cycle-stable and dissolution-resistant (Li et al., 2022).

3.9 Cycle Life Performance

Anthraquinone-based redox polymers' cycle life efficiency affects organic battery life. A battery's cycle life is the number of full charge-discharge cycles before its capacity drops to a certain percentage. The stability of the polymer matrix, the strength of the AQ redox couple, and the battery system's active component integration and control determine the cycle life of AQ-based redox polymers. Actual cycle life may be affected by battery restrictions such as excessive self-discharge, potential imbalance, and more float charge capacity per current (Grieco et al., 2022).

4. Performance in Organic Batteries

A beneficial specific capacity, high coulombic efficiency, excellent rate capability, and possibility of sustainable energy storage solutions of anthraquinone (AQ)-based redox polymers represent an attractive option for use in organic batteries. Poly anthraquinone imides (PAQI), a carbonyl molecule with a high specific capacity, good redox kinetics, and abundant resources, make organic cathode materials for rechargeable Li-ion batteries appear promising (Xin et al., 2024).

4.1 Specific Capacity and Coulombic Efficiency

Charge per mass of active material is known as battery-specific capacity. The capacity of Anthraquinone-based redox polymers depends on anthraquinone's two-electron redox mechanism. Anthraquinone's unit accessibility in polymer matrix impacts capacity. Polymer structure optimization for anthraquinone's unit separation and permeability maximizes anthraquinone's active site application. Capacity is dependent on electrode architecture, including active material size and distribution. The addition of non redox-active inorganic cations (e.g., Na^+) to redox-active organic anions reduces specific capacity (Zhen et al., 2022). A well-designed electrode increases electron and ion movement. Electrolytes alter proton availability and redox reaction dynamics, which affect capacity.

In charge-discharge cycles, coulombic efficiency evaluates charge transfer. The ideal discharge-to-charge ratio is about 100%. The AQ/AH redox couple's reversibility makes AQ-based polymers efficient. This allows efficient charge transfer and few side reactions. After several cycles, polymer matrix and AQ unit stability are needed for high coulombic efficiency. Coulombic efficiency is referred to a specific cycle (Yang et al., 2020). Forming stable redox sites, minimizing polymer backbone degradation, and minimizing electrolyte AQ dissolution maximizes coulombic efficiency. Asymmetric cell architecture with AQDS and iron sulphate electrolytes yielded 99.63% coulombic efficiency and prevented irreversible capacity decline (Yang et al., 2020).

4.2 Rate Capability

Rate capability is known as the battery's ability to deliver capacity at various charge and discharge rates. Active materials' strong conductivity and faster kinetics explain the battery's excellent rate capabilities (Zhen et al., 2021).

4.3 Fast Redox Kinetics

AQ's linked structure accelerates electron transport and protonation/deprotonation. High-rate capabilities are enabled by rapid AQ-AH_2 transitions. The electromagnetic transfer of mass over potential works up to 40% reactant utilisation, which is the amount of reactant changed by the device in a single pass. This is equal to the current split by the product of the reactant concentration and flow rate (Wu et al., 2021).

4.4 Polymer Matrix Design

Conductive polymers or additions improve electrode electron transport and rate performance. A porous polymer structure with a high surface area improves electrolyte penetration and ion diffusion, increasing the rate. After addition, dye molecules absorb UV light for polymerization. The polymerization rate and light intensity energy decrease. This allows liquid crystal molecules to disperse and consolidate, increasing microdroplet and polymer matrix mesh sizes (Zhao et al., 2023).

4.5 Comparative Performance

AQ-based redox polymers surpass inorganic materials. Polymer matrix engineering and molecular design flexibility allow rapid charge-discharge cycles. Stability, specific capacity, and rate capability must exist together. Unsolvated sodium cations are more effectively transported by an electrolyte with a high salt concentration (Chen et al., 2015).

4.6 Stability and Degradation

AQ-based polymers' electrochemical stability during long-term cycling is a major challenge. The polymer backbone or AQ units can degrade, reducing efficacy and capacity. The chemical instability of the reduced form causes capacity loss, but electrochemical responses have little effect (Mazur et al., 2021). AQ or AH solubility in the electrolyte reduces capacity and cycle life. Electrolyte-insoluble AQ derivatives or strong bonds are used to bind AQ to the polymer backbone. AQ or AH dissolve in the electrolyte, limiting battery capacity and charge time (Yang et al., 2020).

4.7 Polymer Processing and Integration

Commercial applications require scalable AQ-based redox polymer production and refining. Production methods must be cost-effective and polymer integration into electrodes must be reliable. Organic materials' structural variability allows for many redox-active polymer topologies for active materials (Hager et al., 2020). AQ-based polymers in electrodes need manufacturing process optimisation for uniform dispersion, mechanical stability, and current collector connection.

4.8 Environmental and Safety Considerations

The environmental impact of AQ-based redox polymer production and disposal is low. The impact of these materials on the environment can be reduced by the use of sustainable synthesis and recycling. Redox-active compounds such as anthraquinone have been used due to their environmental safety aspects, low cost, variety in structure, and high specific capacity (Rasheed et al., 2020). Anthraquinone-based batteries are made more resistant to fire by stabilizing the electrolyte, controlling anthraquinone compounds, limiting overcharge and preventing short circuits. Aqueous redox flow batteries are considered ideal for long-term, storage efficiency, lower environmental impact, and power energy decoupling (Pasadakis-Kavounis et al., 2024).

5. Conclusion

Metal batteries are being replaced by strong and flexible anthraquinone redox polymer batteries. Theoretically, these substances can do strange things and their redox reactions can go both ways. When batteries are put together, their energy, lifespan, and charge cycles go up. They are useful for saving green energy, running portable electronics, and charging electric cars because they have a stable voltage and high coulombic efficiency. Long-term cycle stability and low breakdown of the polymer framework and anthraquinone molecules make the battery more reliable. Using less harmful chemicals when making anthraquinone-based batteries and making recycling better are two ways to make them more environmentally friendly. To make batteries work better, researchers should work on improving redox rates and electrolyte compatibility. Anthraquinone-based redox polymers can be made better with new polymer shapes and production methods. Businesses, universities, and government agencies should all work together to get around scientific problems and bring new battery technology to market. Redox polymers built on anthraquinones may change how energy is stored. Global energy options that are sustainable and work need such new ideas, supported by planned research and development.

References

Akash, M.S.H. and Rehman, K. 2020. Ultraviolet-visible (UV-VIS) spectroscopy. Essentials of Pharmaceutical Analysis, pp. 29–56. https://link.springer.com/chapter/10.1007/978-981-15-1547-7_3.

Amini, K., Kerr, E.F., George, T.Y., Alfaraidi, A.M., Jing, Y., Tsukamoto, T. et al. 2023. An extremely stable, highly soluble monosubstituted anthraquinone for aqueous redox flow batteries. Advanced Functional Materials 33(13): 2211338. https://www.osti.gov/pages/servlets/purl/2349051.

Chen, Jun, Guo, Chunyang, Zhang, Kai, Zhao, Qing and Pei, Longkai. 2015. High-performance sodium battery with 9,10-anthraquinone/CMK-3 cathode and ether-based electrolyte. Chem. Commun. 51. 10.1039/C5CC02251G.https://www.researchgate.net/publication/276366059_Highperformance_sodium_battery_with_910-anthraquinoneCMK-3_cathode_and_ether-based_electrolyte.

Crespo, A. 2021. Use of Anthraquinone Derivatives and Methods of Efficient Syntheses. https://digitalcommons.liberty.edu/cgi/viewcontent.cgi?article=2192&context=honors.

Diaz-Munoz, G., Miranda, I.L., Sartori, S.K., De Rezende, D.C. and Diaz, M.A. 2018. Anthraquinones: an overview. Studies in Natural Products Chemistry 58: 313–338. https://www.sciencedirect.com/science/article/abs/pii/B9780444640567000118.

Dumur, F. 2023. Recent advances in Anthraquinone-based photoinitiators of polymerization. European Polymer Journal 191: 112039. https://www.sciencedirect.com/science/article/am/pii/S0014305723002227.

Fair, R.A., Xie, R., Lee, Y., Colby, R.H. and Gomez, E.D. 2021. Molecular weight characterization of conjugated polymers through gel permeation chromatography and static light scattering. ACS Applied Polymer Materials 3(9): 4572–4578. https://par.nsf.gov/servlets/purl/10302334.

Feng, M., Zeng, R., Chou, S.L. and Gu, F.L. 2021. Enhancing the understanding of the redox properties of lithium-inserted anthraquinone derivatives by regulating molecular structure. Journal of Electroanalytical Chemistry 887: 115172. https://www.sciencedirect.com/science/article/abs/pii/S1572665721001983.

Fontmorin, J.M., Guiheneuf, S., Godet-Bar, T., Floner, D. and Geneste, F. 2022. How anthraquinones can enable aqueous organic redox flow batteries to meet the needs of industrialization. Current Opinion in Colloid & Interface Science 61: 101624. https://www.sciencedirect.com/science/article/abs/pii/S1359029422000632.

Goujon, N., Casado, N., Patil, N., Marcilla, R. and Mecerreyes, D. 2021. Organic batteries are based on just redox polymers. Progress in Polymer Science 122: 101449. https://www.sciencedirect.com/science/article/pii/S0079670021000964.

Grieco, R., Molina, A., Sanchez, J.S., Patil, N., Liras, M. and Marcilla, R. 2022. A significantly improved polymer|| $Ni(OH)_2$ alkaline rechargeable battery using anthraquinone-based conjugated microporous polymer anode. Materials Today Energy 27: 101014. https://www.sciencedirect.com/science/article/pii/S2468606922000727.

Guerrero-Pérez, M.O. and Patience, G.S. 2020. Experimental methods in chemical engineering: Fourier transform infrared spectroscopy—FTIR. The Canadian Journal of Chemical Engineering 98(1): 25–33. https://onlinelibrary.wiley.com/doi/abs/10.1002/cjce.23664.

Hager, M.D., Esser, B., Feng, X., Schuhmann, W., Theato, P. and Schubert, U.S. 2020. Polymer-based batteries—flexible and thin energy storage systems. Advanced Materials 32(39): 2000587. https://onlinelibrary.wiley.com/doi/pdfdirect/10.1002/adma.202000587.

Khan, H., Yerramilli, A.S., D'Oliveira, A., Alford, T.L., Boffito, D.C. and Patience, G.S. 2020. Experimental methods in chemical engineering: X-ray diffraction spectroscopy—XRD. The Canadian Journal of Chemical Engineering 98(6): 1255–1266. https://onlinelibrary.wiley.com/doi/abs/10.1002/cjce.23747.

Kim, J., Kim, Y., Yoo, J., Kwon, G., Ko, Y. and Kang, K. 2023. Organic batteries for a greener rechargeable world. Nature Reviews Materials 8(1): 54–70. https://www.nature.com/articles/s41578-022-00478-1.

Letourneau, D.R. and Volmer, D.A. 2023. Mass spectrometry-based methods for the advanced characterization and structural analysis of lignin: A review. Mass Spectrometry Reviews 42(1): 144–188. https://analyticalsciencejournals.onlinelibrary.wiley.com/doi/pdfdirect/10.1002/mas.21716.

Li, T., Chai, S., Liu, B., Zhao, C. and Li, H. 2023. All-carbon backbone aromatic polymers for proton exchange membranes. Journal of Polymer Science 61(22): 2796–2814. https://onlinelibrary.wiley.com/doi/full/10.1002/pol.20230405.

Lu, X. and Elimelech, M. 2021. Fabrication of desalination membranes by interfacial polymerization: history, current efforts, and future directions. Chemical Society Reviews 50(11): 6290–6307. https://pubs.rsc.org/en/content/articlehtml/2021/xx/d0cs00502a.

Luo, B., Chen, Y., Zhang, Y. and Huo, J. 2021. Nitrogen-rich anthraquinone–triazine conjugated microporous polymer networks as a high-performance supercapacitor. New Journal of Chemistry 45(37): 17278–17286. https://pubs.rsc.org/en/content/articlelanding/2021/nj/d1nj03180e/unauth.

Masek, A., Plota, A., Chrzastowska, J. and Piotrowska, M. 2021. Novel hybrid polymer composites based on anthraquinone and eco-friendly dyes with potential for use in intelligent packaging materials. International Journal of Molecular Sciences 22(22): 12524. https://www.mdpi.com/1422-0067/22/22/12524.

Mazúr, P., Charvát, J., Mrlík, J., Pocedič, J., Akrman, J., Kubáč, L. et al. 2021. Evaluation of electrochemical stability of sulfonated anthraquinone-based acidic electrolyte for redox flow battery application. Molecules 26(9): 2484. https://www.mdpi.com/1420-3049/26/9/2484.

Mohamed, M.G., Ibrahim, M., Ping Chen, N., Basit, A., Chin Kao, Y., Osama Mousa, A. et al. 2024. Tetrabenzonaphthalene and redox-active anthraquinone-linked conjugated microporous polymers as organic electrodes for enhanced energy storage efficiency. ACS Applied Energy Materials. https://pubs.acs.org/doi/epdf/10.1021/acsaem.4c01276.

Nakayama, T., Okumura, N. and Uno, B. 2020. Complementary effect of intra-and intermolecular hydrogen bonds on electron transfer in β-hydroxy-anthraquinone derivatives. The Journal of Physical Chemistry B 124(5): 848–860. https://pubs.acs.org/doi/abs/10.1021/acs.jpcb.9b10733.

National Center for Biotechnology Information. 2024.. PubChem Compound Summary for CID 6780, Anthraquinone. Retrieved July 4, 2024 from https://pubchem.ncbi.nlm.nih.gov/compound/Anthraquinone.

Nawar, S., Huskinson, B. and Aziz, M. 2013. Benzoquinone-hydroquinone couple for flow battery. MRS Online Proceedings Library (OPL) 1491: mrsf12-1491. https://dash.harvard.edu/bitstream/handle/1/27417435/mja233.pdf?sequence=1.

Nguyen-Tri, P., Ghassemi, P., Carriere, P., Nanda, S., Assadi, A.A. and Nguyen, D.D. 2020. Recent applications of advanced atomic force microscopy in polymer science: A review. Polymers 12(5): 1142. https://www.mdpi.com/2073-4360/12/5/1142.

Ntombela, M., Musasa, K. and Leoaneka, M.C. 2022. Power loss minimization and voltage profile improvement by system reconfiguration, DG sizing, and placement. Computation 10(10): 180. https://www.mdpi.com/2079-3197/10/10/180.

Pan, M., Shao, M. and Jin, Z. 2023. Development of organic redox-active materials in aqueous flow batteries: current strategies and future perspectives. SmartMat 4(4): e1198. https://onlinelibrary.wiley.com/doi/full/10.1002/smm2.1198.

Pasadakis-Kavounis, A., Baj, V. and Hjelm, J. 2024. Aggregation and capacity limiting effects in anthraquinone-based flow battery negolytes. Journal of The Electrochemical Society 171(2): 020501. https://iopscience.iop.org/article/10.1149/1945-7111/ad1ec3/pdf.

Qi, N., Yao, B., Sun, H., Gao, Y., Liu, X. and Li, F. 2023. Anthraquinone-based porous organic polymers: From synthesis to applications in electrochemical energy conversion and storage. Arabian Journal of Chemistry, p. 105263. https://www.sciencedirect.com/science/article/pii/S1878535223007256.

Rasheev, H.G., Araujo, R.B., Tadjer, A. and Johansson, P. 2020. The fundamental promise of anthraquinone functionalized graphene-based next-generation battery electrodes: a DFT study. Journal of Materials Chemistry A 8(28): 14152–14161. https://pubs.rsc.org/en/content/articlehtml/2020/ta/d0ta04780e.

Rosemeier, M., Lester, C., Antoniou, A., Fahrenson, C., Manousides, N. and Balzani, C. 2022. Determination of as-built properties of fiber-reinforced polymers in a wind turbine blade using scanning electron and high-resolution X-ray microscopy. Composites Part C: Open Access 9: 100310. https://www.sciencedirect.com/science/article/pii/S2666682022000731.

Schriber, E.A., Paley, D.W., Bolotovsky, R., Rosenberg, D.J., Sierra, R.G., Aquila, A. et al. 2022. Chemical crystallography by serial femtosecond X-ray diffraction. Nature 601(7893): 360–365. https://www.nature.com/articles/s41586-021-04218-3.

Su, Y., Shang, J., Liu, X., Li, J., Pan, Q. and Tang, Y. 2024. Constructing π–π superposition effect of tetralithium naphthalenetetracarboxylate with electron delocalization for robust dual-ion batteries. Angewandte Chemie International Edition, p.e202403775. https://www.researchgate.net/profile/Qingguang-Pan/publication/379258174_Constructing_p-p_Superposition_Effect_of_Tetralithium_Naphthalenetetracarboxylate_with_Electron_Delocalization_for_Robust_Dual-ion_Batteries/links/662b83849d2a69723f68bc08/Constructin.

Wang, Z., Fan, Q., Guo, W., Yang, C. and Fu, Y. 2022. Biredox-ionic anthraquinone-coupled ethylviologen composite enables reversible multielectron redox chemistry for Li-organic batteries. Advanced Science 9(1): 2103632. https://onlinelibrary.wiley.com/doi/pdfdirect/10.1002/advs.202103632.

Wu, M., Bahari, M., Fell, E.M., Gordon, R.G. and Aziz, M.J. 2021. High-performance anthraquinone with potentially low cost for aqueous redox flow batteries. Journal of Materials Chemistry A 9(47): 26709–26716. https://pubs.rsc.org/zh-hans/content/getauthorversionpdf/D1TA08900E.

Wu, M., Jing, Y., Wong, A.A., Fell, E.M., Jin, S., Tang, Z. et al. 2020. Extremely stable anthraquinone negolytes synthesized from common precursors. Chem. 6(6): 1432–1442. https://www.cell.com/chem/fulltext/S2451-9294(20)30134-0?dgcid=raven_jbs_aip_email.

Xie, W., Yu, Z., Huang, H., Jiang, R., Yao, S., Huang, J. et al. 2024. Anthraquinone-based polymer-modified BiVO4 photoanode with strong electron-withdrawing functional groups for boasting photoelectrochemical water oxidation. Journal of Colloid and Interface Science 665: 977–987. https://www.sciencedirect.com/science/article/abs/pii/S0021979724006623.

Xin, Y., Li, Z., Ge, Y., Zhang, F., He, B., Wang, C. et al. 2024. High-rate performance organic cathodes based on carbon nanotube-modified anthraquinone-based polyimides for all-organic lithium/sodium-ion batteries. The Journal of Physical Chemistry C. https://pubs.acs.org/doi/abs/10.1021/acs.jpcc.4c02998.

Yang, B., Murali, A., Nirmalchandar, A., Jayathilake, B., Prakash, G.S. and Narayanan, S.R. 2020. A durable, inexpensive and scalable redox flow battery based on iron sulfate and anthraquinone disulfonic acid. Journal of The Electrochemical Society 167(6): 060520. https://iopscience.iop.org/article/10.1149/1945-7111/ab84f8/pdf.

Yang, J., Su, H., Wang, Z., Sun, P. and Xu, Y. 2020. An insoluble anthraquinone dimer with a near-plane structure as a cathode material for lithium-ion batteries. ChemSusChem 13(9): 2436–2442. https://drive.google.com/file/d/1PTuHy8r2XaqHYPGEei10Q3FfPNyc98Lo/view.

Yin, X., Liu, J. and Jäkle, F. 2021. Electron-deficient conjugated materials via p–π* conjugation with boron: extending monomers to oligomers, macrocycles, and polymers. Chemistry–A European Journal 27(9): 2973–2986. https://chemistry-europe.onlinelibrary.wiley.com/doi/am-pdf/10.1002/chem.202003481.

Zhao, C., Hu, Y., Xu, J., Yu, M., Zou, C., Wang, Q. et al. 2023. Research on the morphology, electro-optical properties and mechanical properties of electrochromic polymer-dispersed liquid crystalline films doped with anthraquinone dyes. Crystals 13(5): 735. https://www.mdpi.com/2073-4352/13/5/735.

Zhen, Y., Zhang, C., Yuan, J. and Li, Y. 2021. Anthraquinone-based electroactive ionic species as stable multi-redox anode active materials for high-performance nonaqueous redox flow batteries. Journal of Materials Chemistry A 9(38): 22056–22063. https://acris.aalto.fi/ws/portalfiles/portal/76171695/CHEM_Zhen_et_al_Anthraquinone_based_electroactive_2021_Journal_of_Materials_Chemistry_A.pdf.

Zhu, D.Y., Deng, Z.H. and Chen, S.W. 2021. A review of nuclear magnetic resonance (NMR) technology applied in the characterization of polymer gels for petroleum reservoir conformance control. Petroleum Science 18(6): 1760–1775. https://www.sciencedirect.com/science/article/pii/S199582262100056X.

Chapter 5

Poly(Ionic Liquids) as Redox Polymers in Energy Storage Applications

Sethupathy Ramanathan,[1] *Panneerselvam Perumal,*[1,*]
Narendra Pal Singh Chauhan,[2] *Sapana Jadoun*[3]
and *Zeinab Nokhbedehghan*[4]

1. Introduction on Ionic Liquids

Paul Walden published the first description of an ionic liquid (ILs) in 1914, observing the unique physical properties of ethyl ammonium nitrate ($[EtNH_3][NO_3]$ (Walden, 1914). At the time, the discovery of this new material did not receive much attention. But it has since been regarded as the foundation for the study of ILs. ILs are salts comprising organic cations and organic/inorganic anions that melt at ambient temperatures (Plechkova and Seddon, 2008). Ionic liquids, composed of ions, exhibit superior ionic conductivity, nonvolatility, and nonflammability. Increasing interest in ionic liquids stems from their unique features. We summarise the distinctive features of ionic liquids in Table 1, as they have been previously described in other works.

[1] Department of Chemistry, Faculty of Engineering and Technology, SRM Institute of Science and Technology, Kattankulathur, Chengalpattu, Tamil Nadu-603203, India.
[2] Department of Chemistry, Faculty of Science, Bhupal Nobles University, Udaipur, Rajasthan, India.
[3] Sol-ARIS, Departamento de Química, Facultad de Ciencias, Universidad de Tarapacá, Avda. General Velásquez, 1775, Arica, Chile.
[4] Department of Medical Biotechnology, Faculty of Allied Medicine, Iran University of Medical Sciences, Tehran, Iran.
* Corresponding author: panneerp1@srmist.edu.in; panneerchem82@gmail.com
DOI: 10.1201/9781003546672-5

Table 1. Basic properties of ionic liquids (ILs).

Low melting point	• Treated as liquid at ambient temperature. • Wide usable temperature range.
Non-volatility	• Thermal stability. • Nonflammability.
Composed by ions	• High ion density. • High ion conductivity.
Organic ions	• Variety. • Designability • Unlimited combination.

The intrinsic nonflammability and absence of volatility in ion-conductive liquids present new opportunities across multiple industries. Nonflammable liquids exhibiting high ionic conductivity are important materials for electrochemical applications. Key properties for these applications include nonvolatility and strong ion conductivity, which are essential for advanced and safe electrolyte solutions, especially in outdoor energy devices. Due to the potential for unforeseen explosions or ignitions in most energy devices, there is significant motivation to pursue safe materials for motor vehicles. Additionally, ILs have been used as solvents for a variety of reactions, including Diels–Alder processes, Suzuki couplings, alkylations, Friedel–Crafts reactions, hydrogenations, metathesis reactions, Heck reactions, and others (Audic et al., 2003; Badri et al., 1992; Chauvin et al., 1996; CJ, 2000; Fischer et al., 1999; Kaufmann et al., 1996; MacLean and Singer, 1999; Rajagopal et al., 2002; Yao and Zhang, 2003). The physical chemistry ideas behind the obvious difference in solution properties between molecular solvents and molten salts have already been discussed and summarised in several articles. Numerous articles have been released, revealing their outstanding qualities and usefulness in a variety of sectors (CJ, 2000; Kaufmann et al., 1996; Rajagopal et al., 2002) .

Ionic liquids (ILs) can be categorised into several types, including room-temperature ionic liquids (RTILs) (Chatel and Macfarlane, 2014; Lei et al., 2014; Mai and Koo, 2016), task-specific ionic liquids (TSILs) (Gurkan et al., 2010; Ruckart et al., 2015), polyionic liquids (PILs) (Qian et al., 2017; Rojas et al., 2014), and supported ionic liquid membranes (SILMs) (Scovazzo et al., 2009; Wickramanayake et al., 2014).

1.1 Polymer Ionic Liquids (or) Poly(Ionic Liquid) (PILs)

Recent progress in polymer science has enabled the development of a wide range of polyelectrolytes, continuously emerging with precisely controlled structures, such as linear polymers, dendritic, nano/microparticles, and star-shaped architectures. A type of polyelectrolytes called PILs have an ionic liquid (IL) part in each repeated monomer unit. These monomers are connected through a polymer backbone, forming a macromolecular structure (Qian et al., 2017; Xu et al., 2018; Yuan et al., 2013). Over the past fifty-five decades, PILs have been extensively studied

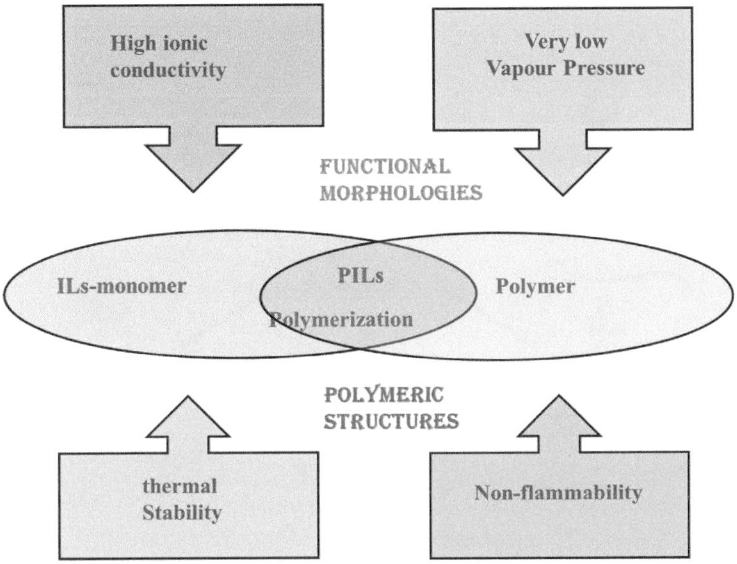

Fig. 1. Principal technique for the synthesis of Poly(ionic liquids) (PILs).

as novel polyelectrolytes, resulting in a significant increase in publications and applications across numerous fields. The features of PILs as shown in Fig. 1, which include excellent ionic conductivity, thermal stability, very low vapour pressure, and non-flammability. These discoveries motivated researchers to employ PILs with varying designs and characteristics in many applications. PILs, as a growing interdisciplinary field spanning polymer chemistry, materials science, physics, separation, catalysis, and electrochemistry, analytical chemistry, are gaining significant attention (Yuan and Antonietti, 2011). Reviews by Firestone, Mecerreyes, Texter, Long, and Antonietti provide comprehensive overviews of the synthesis, characterization, and applications of PILs (Green and Long, 2009; Lu et al., 2009; Mecerreyes, 2011; Shaplov et al., 2015; Yuan and Antonietti, 2011). Research on monomer-based ionic liquids has resulted in the development of innovative solid materials including ionic liquid frameworks. A variety of articles have examined this innovative class of ionic polymers, focusing on the development of new structures, properties, and potential applications (Green and Long, 2009; Lu et al., 2009).

The polymerisation of ionic liquid (IL) monomers is a straightforward and widely utilised method for synthesising polymerised ionic liquids (PILs) across various research groups. Figure 2 presents fundamental forms of IL monomers.

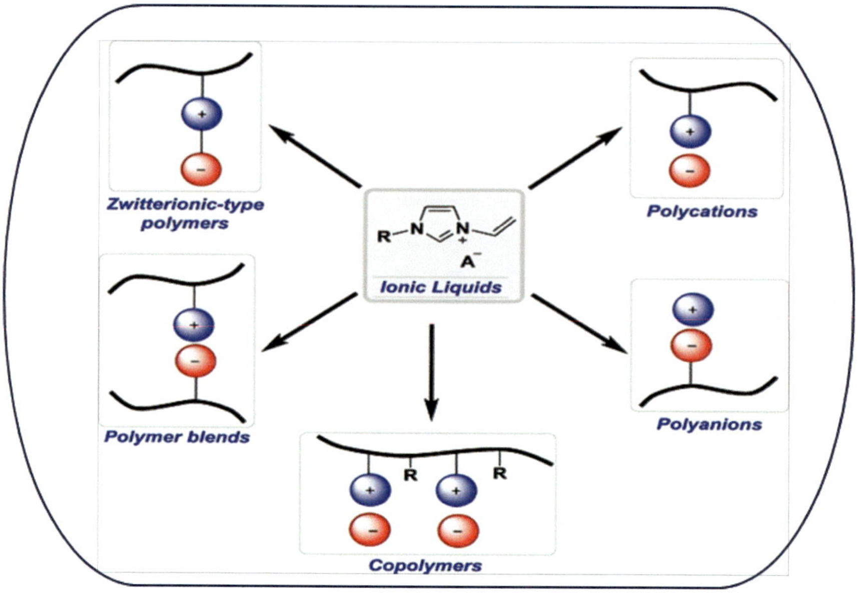

Fig. 2. Fundamental forms of Ionic Liquids (ILs).

2. Synthetic Strategy

2.1 Polymerization of Ionic Liquid

The approach for synthesising PILs consists of the polymerisation of primary ionic liquid monomers (ILMs) adopting with functional cationic species, including imidazolium, pyridinium, or ammonium, accompanied by diverse counter-anions. The characteristics of the final PIL can be modified by changing: **Cation structure** (e.g., imidazolium, pyridinium, ammonium), **Anion type** (e.g., BF_4^-, PF_6^-, Tf_2N^-, Cl^-), **Side chain length** and **functional groups**. This kind of IL monomers have generated significant interest in literature for polymerisation and have frequently been documented. Each reaction in this series comprises the addition of a monomer-ended radical to a monomer. PILs can be synthesized via three basic strategies: (1) direct polymerization of IL monomers with or without non-ionic monomers, (2) Post-polymerization Modification (or) chemical modification of existing polymers, (3) step-growth polymerization (or) Polycondensation. From a synthetic standpoint, each of these approaches, along with the polymerization methods, influences various structural parameters of PILs, presenting unique advantages and limitations in terms of molecular design. The following chapters offer a comparative overview of PIL synthesis, focusing on imidazolium and pyrrolidinium-based PILs, which are among the most widely studied and currently prominent in the field (Salamone et al., 1973).

Direct polymerization of ionic liquid (IL) monomers is based on the characteristics of the polymer backbone, resulting in two types of PILs: polymerized anionic ionic

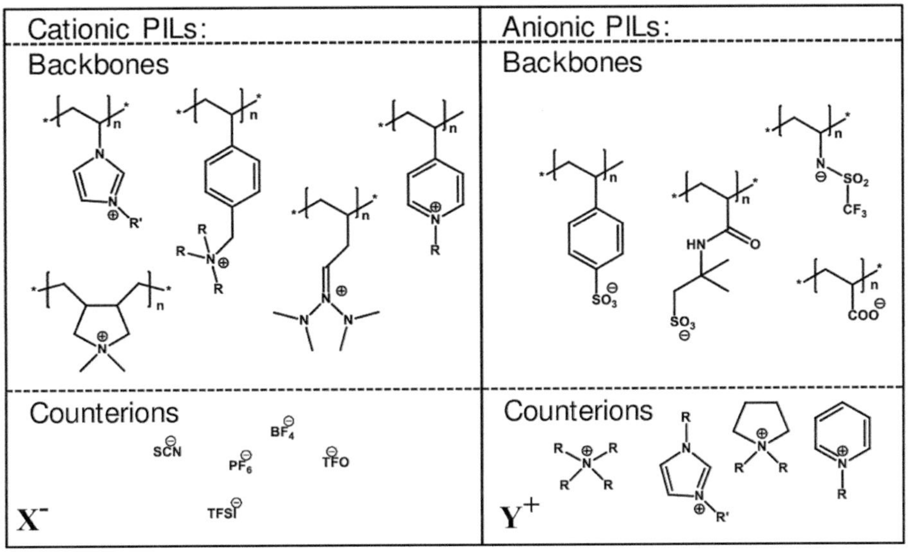

Fig. 3. Illustrations of the chemical structures of cationic, anionic-polymeric ionic liquids (PILs). (x^- is a common counter anion, y^- is a common counter cation, Rs represent alkyl or phenyl group.)

liquids and polymerized cationic ionic liquids. Each type can be synthesized through various methods, including direct polymerization of IL monomers or modifying polymer structures by introducing repeating units that contain IL components. With the growing interest in polymer-based ionic liquids, numerous studies have been published describing various architectures of IL monomers, primarily using acryloyl or vinyl groups, as illustrated in Fig. 3.

In polymerization of IL monomers different mechanisms are involved, most common methods being free radical polymerization, ring opening metathesis polymerization, step-growth polymerization, electrochemical polymerization and many more. These polymerisation mechanisms consist of three step processes, namely chain initiation, chain propagation, and chain termination.

2.1.1 Free Radical Polymerization

Over 60% of polymers are synthesized through free radical polymerization, a technique that, since the 1990s, has been advanced by the Ohno group for producing PILs. By employing various IL monomers, solid ionic conducting PILs has developed by using this method. Their pioneering research has broadened the applications of PILs across multiple fields, positioning traditional free radical polymerization of IL monomers as the leading approach for PILs production (Yuan and Antonietti, 2011). Initiators, such as 2, 20-azo-bis-isobutyronitrile (AIBN), are commonly employed to create free radicals at reaction temperatures. The majority of PILs are derived from imidazolium ions that are functionalised with polymerizable acrylic, styryl, or vinyl groups, resulting in a physical form resembling either a sticky rubber or glass

(Ricks-Laskoski and Snow, 2006). Free radical homo- and copolymerisation have been conducted in ionic liquids.

Ionic liquids based on imidazolium salts have been used to do free radical homopolymerizations of methyl methacrylate (MMA) and glycidyl methacrylate (GMA). The propagation rate coefficients (kp) in IL solutions grew by as much as fourfold for MMA and twice for GMA relative to bulk polymerization conditions (Lu et al., 2009). Salamone and colleagues investigated the homopolymerization of 1-vinyl-3-alkylimidazolium iodide with varying alkyl chain lengths, discovering that the solubility of the resulting polymer solutions was affected by the length of the alkyl chains (Salamone et al., 1973). Recent studies indicate that free radical polymerization in ionic liquids is significantly influenced by the liquids' form, viscosity, and polarity. In particular, the higher viscosity of ionic liquids tends to enhance the degree of polymerization in the resulting polymers. Copolymers of PILs enable the incorporation of IL species into polymer structures either selectively or in a random distribution through the copolymerization process (Bezdushna and Ritter, 2009). Mu et al. demonstrated this by copolymerizing N-vinyl-2-pyrrolidone with 1-vinyl-3-butylimidazolium halide as the initiator using AIBN in methanol at 60°C, as shown in Fig. 4 (Mu et al., 2005).

The polymerised ionic liquid (Poly(MEBIm-BF4)) described by Chen and Elabd (Chen and Elabd, 2009) was made using standard free radical polymerisation (Fig. 5 scheme). The monomer was prepared by quaternizing 2-bromoethyl methacrylate with 1-butylimidazole, followed by an anion exchange reaction. High conductivities were attained by the solid-state fibre mats produced by electrospinning the polymer solutions.

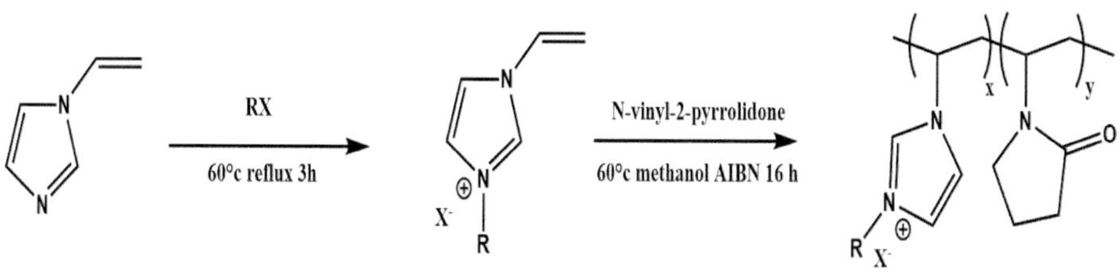

Fig. 4. Synthesis of Poly(NVP-VAH) by copolymerization.

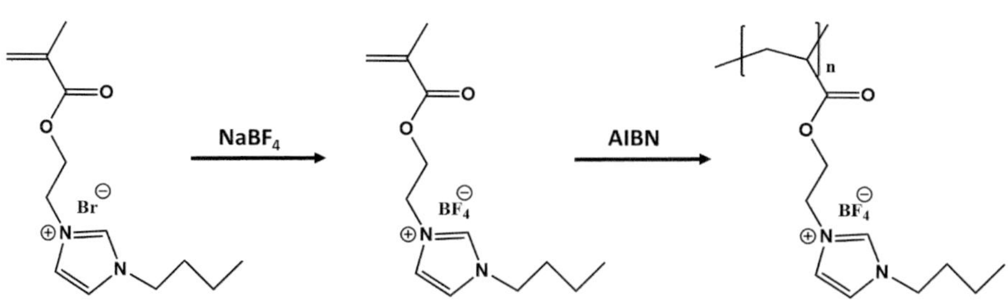

Fig. 5. Preparation of poly[MEBIm-BF_4].

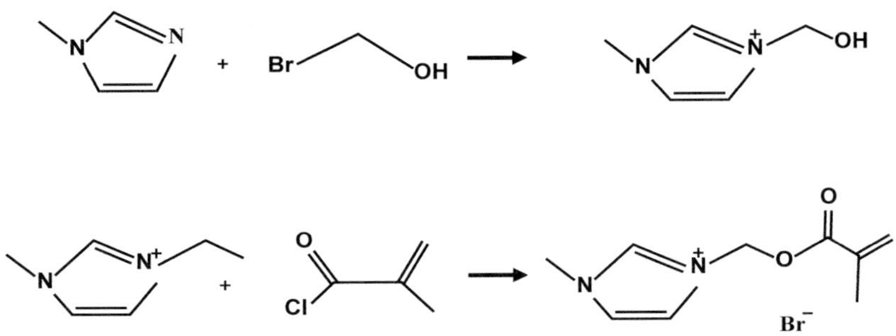

Fig. 6. Synthesis of 1-(n-hydroxyalkyl)-3-methylimidazolium bromides.

N-vinylimidazolium-based IL monomers can readily undergo free radical polymerization in the presence of a thermal initiator (Fig. 6). An ionic liquid monomer containing a tetrafluoroborate anion was made by mixing N-vinyl imidazole with tetrafluoroboric acid in a water-ethanol solvent. Silver salts can help with the anion exchange process for anions like nitrate and dicyanamide that have little effect on the hydrophobicity of IL monomers (Hirao et al., 2000).

Polymerizable ionic liquids (PILs) with methacrylate groups and imidazolium cations have been developed with spacer lengths between the polymerizable functional group ranging from 2 to 10 carbon atoms. A series of 1-(n-hydroxyalkyl)-3-methylimidazolium bromides and 1-[n-(methacryloyloxy)alkyl]-3-methylimidazolium bromides were synthesized, achieving high yields (Pohako-Esko et al., 2013).

The free-radical polymerization of methyl methacrylate and styrene in room-temperature ionic liquid 1-butyl-3-methylimidazolium hexafluorophosphate ([C4mim][PF6]) produces polymers with strengths up to 100 times greater than those obtained with benzene. This is achieved without the use of volatile organic compounds (VOCs). This approach not only enhances polymer strength but also offers significant economic and environmental advantages (MatthewáReichert, 2002).

In the past two decades, controlled/living radical polymerisation technologies have rapidly advanced, affecting polymer science and PILs. Recent breakthroughs such as these allow meso/nanoscale construction and regulation of IL species macromolecular architecture in polymers. Here, these efforts were achieved by two new methods, such as Atom transfer radical polymerization (ATRP) and reversible addition fragmentation chain transfer (RAFT) polymerisation to produce PIL homopolymers and block copolymers. By using ATRP method, namely 1-(4-vinylbenzyl)-3-butyl imidazolium tetrafluoroborate (VBIT) and 1-(4-vinylbenzyl) 3-butylimidazolium hexafluorophosphate (VBIH) was synthesis from styrenic ionic liquid monomers (Fig. 7) (Tang et al., 2005a). Cu(I)Br/2,20-bipyridine catalyst system with ethyl 2-bromoisobutyrate initiator resulted in well controlled polymerization. In the RAFT polymerization of methacryloyl-based IL monomers, developed by Gnanou and Shen et al., an alkyl spacer is used to separate the imidazolium cation from the polymerizing unit. The nonconjugated nature of N-vinylimidazolium-based IL

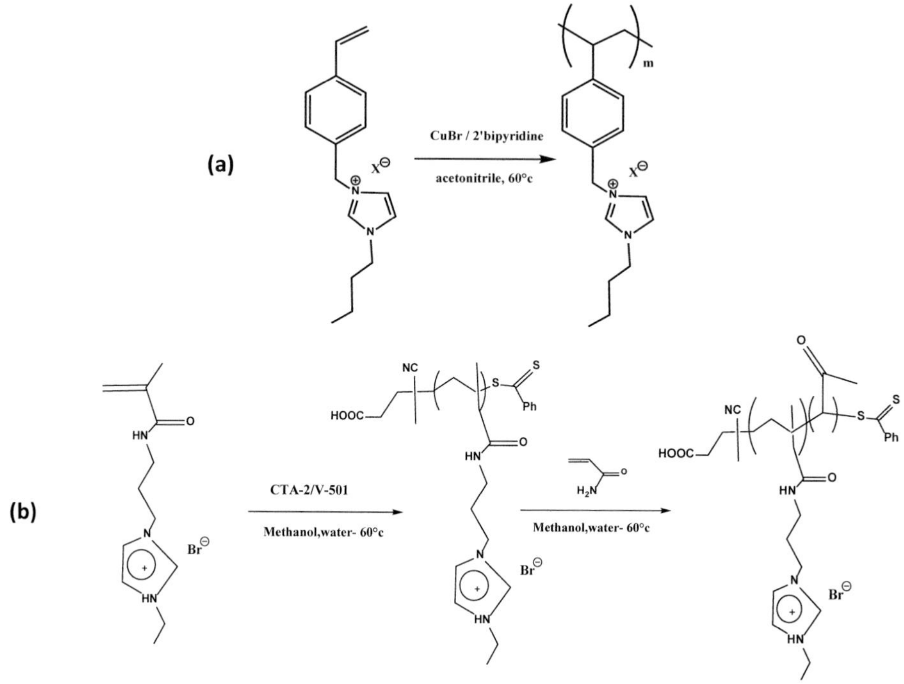

Fig. 7. (a) Synthesis of 1-butyl-3-vinylbenzylimidazolium tetrafluoroborate or hexafluorophosphate in acetonitrile by ATRP (X: BF_4^- or PF_6^-) and (b) Double-hydrophilic deblock copolymers PILs by RAFT polymerization.

monomers complicates controlled radical polymerization due to the high reactivity of the radical species involved (Vijayakrishna et al., 2008; Yang et al., 2008).

2.1.2 Post-Modification Polymerization

Post-modification is a widely used strategy for synthesizing PILs (Carrasco et al., 2011; Korolovych et al., 2018; Obadia et al., 2015; Stancik et al., 2004; Wang et al., 2017). Common groups employed for this modification include benzyl bromide, benzyl chloride, and various alkaline groups, such as pyridine and amines, which introduce ionic moieties into the polymer chain. In general, the post-modification process involves nucleophilic addition reactions. C.M. Stanick et al. described a method for synthesizing PILs through this approach (Fig. 8) (Stancik et al., 2004). In their study, a benzyl chloride-functionalized polymer was allowed to react with methylimidazole, followed by anion exchange to produce a PIL with BF_4^- as the counterion.

Wang et al. presented the PILs synthesized using a post-modification technique involving a nucleophilic addition reaction between a benzyl chloride-based polymer chain and a neutral imidazole-based polymer chain (Wang et al., 2017). Their method involved the copolymerization of 1-(Chloromethyl)-4-vinylbenzene and divinylbenzene (DVB) through conventional free radical polymerization, with imidazole followed by nucleophilic addition to produce the final polymeric ionic

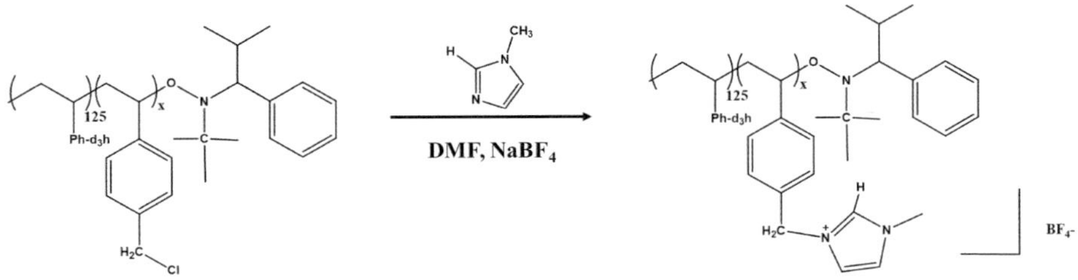

Fig. 8. Synthesis of PIL via post-modification strategy.

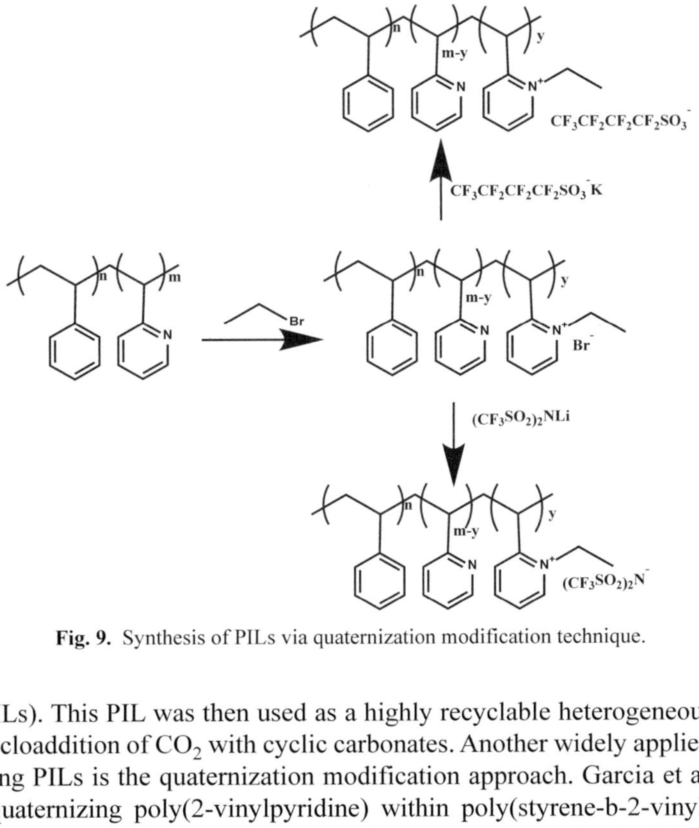

Fig. 9. Synthesis of PILs via quaternization modification technique.

liquid (PILs). This PIL was then used as a highly recyclable heterogeneous catalyst for the cycloaddition of CO_2 with cyclic carbonates. Another widely applied strategy for creating PILs is the quaternization modification approach. Garcia et al. applied this by quaternizing poly(2-vinylpyridine) within poly(styrene-b-2-vinylpyridine) (PS-b-P2VP) via nucleophilic addition with ethyl bromide (Fig. 9) (Obadia et al., 2015). The resulting PIL, could be modified into other PILs through anion exchange processes. Atomic force microscopy (AFM) was used to examine the effect of various anions on the nanostructures of the PILs.

The polymerizable choline-based ionic liquid [2-(methacryloyloxy)ethyl]-trimethylammonium (TMAMA/Cl^-) was functionalised with antibacterial agents cloxacillin (CLX^-) and fusidate (FUS^-) by an ion exchange process. For drug delivery applications, MMA was used to copolymerize biocompatible IL monomers

TMAMA/FUS⁻ and TMAMA/CLX⁻ to produce graft copolymers with 19–50 mol% TMAMA units. In PBS medium, pharmaceutical anions and phosphates exchanged for 53% of FUS⁻ (3.6 μg/mL) and 44% of CLX⁻ (2.7 μg/mL) in individual systems. The pharmaceutical anion type significantly influences the distribution of the polymer conjugate system (Mazur et al., 2022).

2.1.3 Step-Growth Polymerization

In order to achieve large conversion rates and substantial molecular weights, step-growth polymerizations are often carried out in non-solvent or low-solvent environments due to their sluggish reaction kinetics and requirement for high temperatures. Nonetheless, polyimides and polyamides, recognized for their remarkable thermal and mechanical characteristics, are occasionally synthesized in a solution under elevated temperature conditions. Vygodskii and colleagues (Vygodskii et al., 2002) explored step-growth polymerization in various ionic liquids, successfully synthesizing high-molecular-weight polyimides and polyamides without the need for a catalyst. Since then, ionic liquids (ILs) have proven to be highly effective as both solvents and catalysts for a wide range of step-growth polymerizations, including those for polyesters, polyamides, polyethers, poly(ether ketones), polyurethanes, poly(ether sulfones), poly(amino acids), and poly(amide imides) (Dou et al., 2012; Gunaratne et al., 2013; Kanetaka et al., 2016; Mallakpour and Dinari, 2013; Rafiemanzelat and Abdollahi, 2012; Taghavi et al., 2013; Wang and Liu, 2013; Wang et al., 2016; Zhang et al., 2011).

Recently, a sustainable approach for producing high-performance thin-film composite aromatic polyamide membranes using ionic liquids as the reaction medium was introduced by Marien et al. (Marien et al., 2016) described the production of poly[(1-butyl imidazolium-3)methyl] ethylene oxide tetrafluoroborate (p[BIEO][BF_4]) involving modifying polyethylene oxide. The poly(ethylene oxide) backbone is synthesized from poly(epichlorohydrin) through the incorporation of N-butylimidazole, aluminium isopropoxide, and sodium tetrafluoroborate ($NaBF_4$) (Fig. 10) (Tang et al., 2005b).

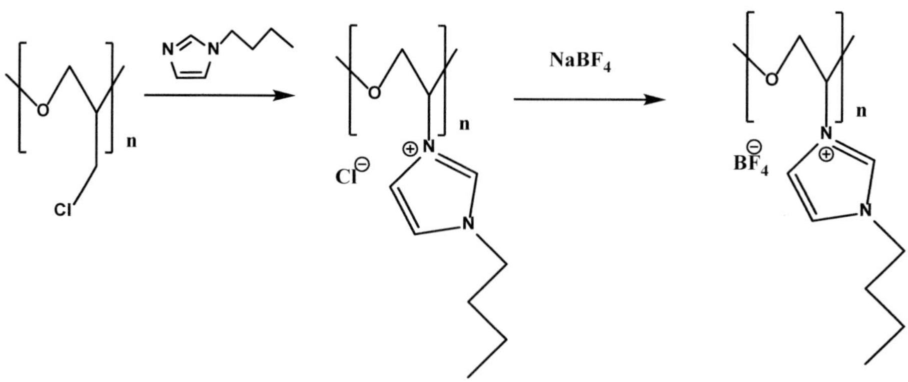

Fig. 10. Synthesis of p[BIEO][BF_4^-].

Literature extensively documents PILs derived from polybenzimidazoles (PBIs). The synthesis of these PBIs was achieved via a polycondensation process, utilizing polyphosphoric acid (PPA) as the dual agent for both condensation and solvent functions. Furthermore, the methodical alteration of the PBI family, which includes a tetraamine (3,3-diaminobenzidine) alongside a range of carboxylic acids, has been investigated in additional research. Furthermore, the synthesis of functional imidazolium-based ionic liquids containing hydroxyl groups and symmetrical ester has been explored. Following condensation polymerization, these monomers were utilized to create PILs, with stannous octoate acting as a catalyst at comparable molar concentrations (Bhavsar et al., 2014; Kumbharkar et al., 2014).

2.1.4 Ring-Opening Polymerization

Ring-opening polymerisation (ROP) in ionic liquids (ILs) has been a prominent research focus due to its capacity to enable polymer synthesis under moderate circumstances while allowing for control over polymer design. The unique features of ionic liquids, including their variable viscosity and thermal stability, improve the efficacy of ring-opening polymerisation processes, rendering them appropriate for diverse applications. Gin et al. recently used ring opening metathesis polymerisation (ROMP) to produce alkylimidazolium-based PIL block and norbornene copolymers with the help of a Grubbs' first-generation catalyst (Wiesenauer et al., 2011). Subsequent studies by this group led to the development of the first ABC PIL triblock copolymers, which were constructed from norbornene backbones and featured alternating hydrophilic, and hydrophobic substituents (Wiesenauer et al., 2013). A highly successful ring-opening polymerization of norbornene has been carried out in a two-phase medium comprising butyl-2,3-dimethylimidazolium hexafluorophosphate ([BdMIM]pF6) and toluene. In this system, the catalyst is located in the ionic liquid (IL) phase, while the toluene phase aids in breaking down the polymer product. Remarkably, both the ionic liquid and catalyst were recyclable up to six times, consistently delivering the same high performance with each cycle as observed in the first use (Csihony et al., 2002). Nishide and colleagues subsequently synthesized a norbornene-based PIL block copolymer with a redox-active segment and an ethyl imidazolium-containing PIL segment, using ring-opening metathesis polymerization (ROMP) with a third-generation Grubbs' catalyst (Suga et al., 2014). Ring-opening polymerisation in ionic liquids (ILs) involves the synthesis of aliphatic polyesters, including copolymers of rac-lactide and ε-caprolactone, facilitated by lipase. Ionic liquids augment polymerizable, leading to higher molecular weights and varied microstructures appropriate for biological applications (Piotrowska et al., 2017). Njoroge, 2017 discuss surface-initiated ring-opening metathesis polymerisation (SI-ROMP) of ionic liquid-tethered monomers, focussing on the development of poly(ionic liquid) films. The procedure facilitates swift film development and anion exchange, improving the versatility of the resultant poly(ionic liquid) films. Effective anion exchange adjusts film characteristics and speeds up the film growth to 600 nm in 15 minutes (Njoroge et al., 2017). The enzymatic ring-opening polymerisation (ROP) of lactides in ionic liquids, namely [BMIM][PF6], can produce polylactides with molecular weights about 20,000 with modest yields. Essential parameters include lipase type, solvent concentration, water content, and reaction temperature (Zhao et al., 2017).

3. Characterization Methods of PILs

The characterization techniques for PILs differ significantly from those used for conventional molecular ionic liquids (ILs). For example, the molecular weight (MW) of PILs is typically expressed as a statistical average, often determined through gel permeation chromatography (GPC). In contrast, the MW of small molecular ILs is an exact value, measured using electrospray ionization-mass spectrometry (ESI-MS). High-resolution nuclear magnetic resonance (NMR) spectroscopy is commonly used to analyze the structure of soluble PILs in liquid form, while solid-state NMR, despite its potentially lower resolution, is necessary for analyzing insoluble PILs. Fourier transform infrared (FTIR) spectroscopy is an effective tool for identifying specific bond structures and functional groups within PILs (Xu et al., 2018; Yuan and Antonietti, 2011; Yuan et al., 2013). Thermogravimetric analysis (TGA) is used to evaluate the thermal stability of PILs, while differential scanning calorimetry (DSC) is employed to measure their glass transition temperature. The specific surface area and pore architecture of porous PILs can be evaluated using nitrogen physisorption techniques. Additionally, high resolution scanning electron microscopy (HRSEM) and transmission electron microscopy (HRTEM) offer direct visualization of structural features such as pore and crystalline structures.

4. Essential Properties of PILs for Energy Applications

Poly(ionic liquids) (PILs) are polymers that integrate ionic liquid (IL) moieties into their structure, providing them with a variety of distinctive and adjustable unique features. They are currently used as significant ionogel electrolytes in solid-state electronics. Recent investigation were conducted to elucidate the inherent features of PILs (ionic conductivity, hydrophilicity, hydrophobicity, glass transition temperature, breakdown, etc.). PILs possess a unique mixture of properties originating from both ionic liquids and polymers, rendering them adaptable materials for many applications. Their structural attributes, mechanical quality, and functional capacities are affected by elements like ion size, density, and the form of ionic interactions. The structural properties of PILs have a multi-scale architecture shaped by ion dimensions and Coulomb interactions, resulting in unique polymer arrangement and charge separation effects. The existence of both rigid and flexible segments in PILs improves their ionic conductivity and electrochemical stability, which are essential to use in solid-state batteries (Chang et al., 2024; Das et al., 2024). PILs have superior mechanical properties such as flexibility and durability, which are crucial for membrane applications. The macromolecular structure improves their thermal stability, enabling high-performance applications in a variety of conditions (Sun et al., 2024). The study examines poly(3,4-ethylenedioxythiophene) (PEDOT) in imidazolium ionic liquids, emphasising their moisture stability, raised n-doping levels, and improved ion exchange rates during redox switching, which enhance the electrochemical characteristics of conducting polymers for diverse applications.

PEDOT sheets exhibit strong n-doping in imidazolium ionic liquids, with varying conductivities (Sandoval et al., 2014).

The cation's composition significantly influences the material by ionic conductivity. Changing the cation from ammonium to imidazolium or pyrrolidinium leads to significant differences in conductivity (10^{-7} S cm^{-1}, 3×10^{-7} S cm^{-1}, and 3×10^{-6} S cm^{-1}, respectively) (Chen et al., 2009). Additionally, the decreasing length significantly influences conductivity. Research indicates that reducing the distance between the cation and the polymer backbone leads to decreased conductivity. For example, in poly(imidazolium)-based ionic liquids, the conductivity increases from 2.4×10^{-10} S cm^{-1} to 3.2×10^{-6} S cm^{-1} as the spacer length changes from ethyl to undecyl. For anionic particles, achieving high conductivity requires the spacer to be as short as possible while using the same cation ($EtIm^+$). This trend suggests that conductivity generally increases with a longer spacer in cationic polymers but decreases with a shorter spacer in anionic polymers. Adding an ionic liquid to a polymer matrix increases the conductivity of the electrolytes in the polymer matrix. The ionic conductivities for the IL-, poly[BVIm][Tf2n]/IL-, and poly[BVIm][Tf2n]/IL mixture were 6.76×10^{-3} S cm^{-1}, 2.92×10^{-3} S cm^{-1}, and 5.83×10^{-3} S cm^{-1}, respectively (Chen et al., 2012). Gu et al. recently presented a polymerizable ionic liquid, consisting of both cations and anions. Their research indicated that the stimuli-responsive characteristics of imidazolium might substantially affect proton conductivity (Gu et al., 2016). A research study by Elabd's group found that replacing the BF_4^- anion with the bis(trifluoromethanesulfonyl)imide ($TFSI^-$) anion improved the ionic conductivity of PILs (Chen et al., 2009).

Alongside the ionic conductivity, the thermal properties of the material play a crucial role in making PILs appealing for a range of applications. A decreased glass transition temperature (Tg) facilitates an increase in ionic conductivity. PILs possess counterions that are loosely bound, larger in size, and exhibit lower mobility compared to those typically found in conventional linear polyelectrolytes. Due to the weak electrostatic interactions between ion pairs, PILs can exhibit a low glass transition temperature (Tg) even with a high charge density (Green et al., 2009). Polymers featuring a poly(1-butyl-3-vinylbenzylimidazolium) (P[VBBIM]) backbone, categorised by varying anions and arranged in order of increasing Tg, are as follows: Poly[VBBIM][Tf_2N] (31°C) << Poly[VBBIM][Sac] (40°C) << Poly[VBBIM][BF_4] (78°C) << Poly[VBBIM][PF_6] (85°C). The organic Tf_2N^- and o-benzoic sulphimide anions (Sac^-) have the capability to significantly reduce the glass transition temperature (Tg) of the polymers due to their plasticising effect on the bulk polymer (Tang et al., 2005b). Anion chemical structures have an impact on PILs T_g set values as well. A comparative study investigated how different counter-anions affect the thermal stability of poly(1-vinyl-3-ethyl-imidazolium) X- PILs. The order of thermal stability among various anions, X-, was observed as follows: $CF_3SO_3^-$ > $(CF_3SO_2)_2N^-$ > $C_{12}H_{25}C_6H_4SO_3^-$ > PF_6^- > Br^- > $C_{16}H_{34}PO_4^-$. Similar trends in thermal stability have been observed in other PIL families and in comparable ionic liquid series (Marcilla et al., 2005).

The molecular weight of poly(ionic liquid)s (PILs) plays a crucial role in influencing their glass transition temperature (Tg), viscosity, modulus, and ionic conductivity. Studies show that for imidazolium-based PILs, ionic conductivity decreases as molecular weight increases, reaching an optimal value around ~ 105 Da. When moving from monomers to trimers, conductivity drops significantly, though beyond ~ 1000 Da, it remains relatively stable. PILs having molecular weights between 480 and 160,400 Da and polymerization degrees ranging from 1 to 333 were studied by Fan et al. Their findings indicated that with an increase in molecular weight, the dynamics of ion transport transitioned from a strong reliance on segmental motion to a state that was predominantly independent of it (Fan et al., 2016).

5. PILs in Energy Applications

PILs have emerged as versatile materials with a wide range of applications due to their unique properties, combining the characteristics of ionic liquids and polymers. Their multifunctionality allows them to be utilized in various fields, including lubrication, energy storage, membrane technology, and sensing. In this chapter, we specifically focus on electrical energy storage applications. PILs have emerged as advantageous materials in energy storage owing to their peculiar ionic characteristics, thermal stability, and structural adaptability. By integrating the benefits of ionic liquids (ILs) with polymer backbones, PILs provide a stable structure capable of housing mobile ions, which is crucial for high-performance energy storage systems. Their adaptability renders PILs appropriate for many applications, encompassing batteries, supercapacitors, and fuel cells. This document provides a summary of the function of PILs in energy storage, covering significant applications, advantages, problems, and current developments.

5.1 Electrochemical Supercapacitors

PILs are gaining particular attention as important elements in the advancement of supercapacitors, thanks to their distinctive characteristics that improve performance indicators like energy density, conductivity, and mechanical stability. In contrast to traditional liquid electrolytes that frequently pose risks of flammability and volatility, PILs demonstrate high ionic conductivity, non-flammability, and thermal stability. Operating at a wider electrochemical potential window (up to 5–6 V), they avoid the hazards related to conventional electrolytes. Over the last ten years, PILs have found applications in high-performance supercapacitors (Ray and Saruhan, 2021). The incorporation of PILs into supercapacitor designs has resulted in novel electrolyte formulations that deal with prevalent issues such as self-discharge rates and operational flexibility. Moreover, PILs can be tailored for particular features, increasing their compatibility with electrodes and enhancing overall device performance (Taghavikish et al., 2018).

Bilayer heterogeneous PILs electrolytes (BHPE) containing anionic and cationic PILs are used to create high-performance capacitors with an ultralow self-discharge rate. A series of BHPE with cationic poly(ionic liquid)s (PILs) featuring a long hexyl alkyl side chain and a small BF_4^- anion exhibits an exceptionally a high energy density of 34.8 Wh kg^{-1} and slow self-discharge rate of –0.04 V h^{-1}, indicating significant potential for further practical applications (Dong et al., 2024). Protic 1-vinylimidazolium hydrogen sulfate and bis-vinylimidazolium hydrogen sulfate are copolymerized *in situ* in a medium of protic 1-methylimidazolium hydrogen sulfate to create the poly(ionogel) effectively. This poly(ionogel) exhibits a high conductivity of 14 mS cm^{-1} at 140°C, attributed to the shared protic ionic species between the poly(ionic liquid) network and the ionic liquid. Supercapacitors made with this ionogel, using polyaniline and carbon nanotube electrodes, achieve a peak specific capacitance of 1132 mF cm^{-2} and maintain 96% of their capacitance after 10,000 charge-discharge cycles at 140°C (Wang et al., 2023). The poly(ionic liquid)/polyacrylamide double-network (PIL/PAM DN) hydrogel electrolytes, synthesized via a two-step γ-radiation technique and infused with diverse metal salts, demonstrate remarkable efficacy. The symmetric SC-Li_2SO_4 cell notably attains an extensive operating voltage range reaching 1.5 V, a maximum energy density of 26.0 Wh·kg^{-1}, and exhibits an impressive 94.1% capacitance retention following 10,000 charge-discharge cycles. The observed stability arises from the deactivation of free water molecules, facilitated by the synergistic interactions among PIL, PAM, and SO_4^{2-} ions (Wei et al., 2024). Light-induced polymerisation was employed under UV irradiation to synthesise the [C_4mIM]Cl/chitosan/pHEMA gel polymer electrolyte, which exhibits exceptional capacitance retention after 2000 charge-discharge cycles at 100°C with a constant current density of 0.5 A g^{-1} (Liu et al., 2014). The poly(vinylidene fluoride-co-hexafluoropropylene)/poly(ethylene glycol) diacrylate P(VDF-HFP)/PEGDA/IL-L film undergone extensive testing, exceeding 1000 charge and discharge cycles. Polymer matrix containing immobilised various ionic liquids (in solid or liquid aggregate form) serving as both separator and electrolyte in a supercapacitor cell (Novakov et al., 2023). Kim et al. and colleagues explored the energy storage capabilities of a graphene electrode modified with poly(1-ethyl-3-vinylimidazolium) Tf_2N^-. In this hybrid structure, the PIL attaches to reduced graphene oxide (rGO) through non-covalent cation–π interactions, resulting in improved energy storage performance (Kim et al., 2011). A poly(propylsulfonate dimethylammoniumpropylmethacrylamide) PIL gel electrolyte, applied to a flexible graphene electrode, demonstrated exceptional performance delivering a high capacitance of 300.8 F cm^{-3} at 0.8 A cm^{-3} and the durability to withstand 10,000 charge-discharge cycles (Peng et al., 2016). Marcilla et al. formulated an electrolyte using poly(diallyldimethylammonium) bis(trifluoromethanesulfonyl)imide within the ionic liquid N-butyl-N-methylpyrrolidinium bis(trifluoromethylsulfonyl)imide ([PYR14][TFSI]), which was subsequently employed in solid-state supercapacitors (Tiruye et al., 2015). Subsequent research with pyrrolidinium-based polymer ionic liquid electrolytes with increased ionic conductivity markedly enhanced the specific capacitance.

5.2 Li-batteries

PILs are increasingly recognized as electrolytes in lithium-ion batteries owing to their distinctive attributes, including enhanced ionic conductivity, thermal stability, and safety. Li-ion batteries are the most commonly used and widely adopted battery type globally. Polymer solutions have been used in the past owing to the fact that the technology is capable of performing a wide variety of functions. They may be developed into many electrolyte types, including liquid, quasi-solid, and solid forms, hence optimizing battery performance by increasing energy density and mitigating flammability issues linked to conventional organic solvents (Ma et al., 2023; Zhang et al., 2020). Recent developments encompass zwitterionic polymer ionic liquids that demonstrate elevated lithium-ion transference numbers and superior cycle stability, rendering them advantageous for next-generation battery technology (Liu et al., 2023). In 2009, researchers Mecerreyes et al. and Appetecchi et al. reported PIL-based gel electrolytes using a poly(diallyldimethylammonium) backbone. The resulting PIL demonstrated substantial chemical stability when in contact with the lithium anode, along with excellent room-temperature conductivity and an expansive electrochemical stability window of approximately 7.0 V. Preliminary battery tests suggested that Li/$LiFePO_4$ solid-state cells could deliver around 140 $mA{\cdot}h{\cdot}g^{-1}$ at 40°C (Appetecchi et al., 2010; Pont et al., 2009). Yin and his team developed a dicationic imidazolium-based poly(ionic liquid), poly(N,N,N-trimethyl-N-(1-vinylimidazolium-3-ethyl)-ammonium bis(trifluoromethanesulfonyl)imide) (Yin et al., 2015). This electrolyte exhibited a low glass transition temperature (~ 54°C), outstanding thermal stability up to 330°C, and notable ionic conductivity of around 10^{-4} $S{\cdot}cm^{-1}$ at moderate temperatures. These features enhance battery performance, highlighting the significant potential of electrolytes based on dicationic PIL for lithium metal batteries. Pyrrolidinium-based poly(diallyldimethylammonium)-bis(trifluoromethane sulfonyl)imide (PDADMA-TFSI) poly(ionic liquids) containing fluorinated anions (FSI, TFSI, BETI, and CFSO) have been utilized for the first time as cathode binders in Li-ion and Li-air batteries. High-voltage NMC 532 cathodes utilizing fluorinated PDADMA binders demonstrated enhanced cell performance, characterized by increased capacity, superior rate performance, and improved cycle stability under accelerated aging conditions. The PDADMA-CFSO binder in cathodes resulted in a 26% increase in cell capacity at 5C (12-minute charge) compared to the PVDF binder. In solid-state Li-O_2 cells, using PDADMA-CFSO as a binder instead of traditional lithiated Nafion improved Coulombic efficiency by 146% and cycling capacity by 70% (Vauthier et al., 2023). Li et al. developed a flexible polymer electrolyte film utilizing a lithium-containing zwitterionic polymer ionic liquid (Lu et al., 2017). By dissociating Li+, the zwitterionic effect can increase the mobility of lithium ions in polymer electrolytes. Using LiTFSI/IL as the ion-conducting phase, Zhang et al. presented a high-performance PIL/silica hybrid ionogel electrolyte (Li et al., 2016). For over 500 hours (at 0.1 $mA{\cdot}cm^{-2}$), this electrolyte maintained consistent cycling of the Li symmetric cell, promoting uniform Li electrodeposition and inhibiting the development and growth of lithium dendrites. Kang et al. described a hybrid electrolyte developed through an *in situ* synthetic approach (Zhou et al., 2017). The hierarchical PIL-based solid electrolyte (HPILSE) was composed of a PDDATFSI membrane and a poly(C1-4TFSI) crosslinked network. At 25°C, the LiFePO4/Li-HPILSE/Li cell exhibited an impressive specific capacitance of 162.4 $mA{\cdot}h{\cdot}g^{-1}$, maintaining a stable electrochemical window of 4.4 V. Additionally, it demonstrated excellent interfacial contact between the electrolyte and the electrode, ensuring optimal performance. In order to create effective ion-conducting channels that improve ionic conductivity, Sato et al. reported using surface-initiated atom transfer radical polymerization (ATRP) to graft poly(N,N-diethyl-N-(2-methacryloylethyl)-N-methylammoniumTf2n) onto monodisperse silica particles (SiO_2 NPs with a diameter of 130 nm) (Sato et al., 2011). The PIL-based sulfur cathode demonstrates exceptional rate performance and remarkable cycling durability, characterized by an ultralow capacity fading rate of 0.062% per cycle over 800 cycles at 2 C. Despite a high sulfur loading of 9.27 mg cm^{-2}, an capacity per unit area of 8.71 mAh cm^{-2} can be attained (Lin et al., 2024).

5.3 Fuel Cells

Fuel cells are an innovative clean energy technology that convert the chemical energy of fuels into electrical energy, while producing no harmful pollutants. Nafion, a commonly used polymer electrolyte membrane, is known for its excellent mechanical and thermal properties; it maintains high conductivity even when hydrated fuel cells face performance challenges at temperatures above 80°C because of water evaporation and reduced water affinity, which hinder their efficiency. To overcome this, replacing water with low-volatility, nonaqueous proton carriers such as protic ionic liquids (ILs) offers substantial advantages, allowing fuel cells to function efficiently at higher temperatures under anhydrous conditions. The Poly([HSO_3BVIm][TfO]$^-$-10) membrane exhibits exceptional proton conductivity, ranging from 0.209 to 0.293 S/cm at 25–40°C, surpassing the performance of the commonly used Nafion membrane, even at lower temperatures. At 160°C, a 10% OPBI/PVImBr(H)-g-SiNP solution attains a proton conductivity of 0.25 S/cm, even reaching this value at 170°C. Although the OPBI/PVImBr(H)-g-SiNP-10% membrane has relatively low mechanical strength at 2.76 MPa, it performs effectively under high-temperature conditions (Zunita et al., 2022). To develop a self-standing polymer electrolyte membrane, a hydrophobic styrene chain segment is strategically incorporated between the poly(ionic liquids) of poly(vinylphosphonic acid/1-propylimidazole) (P(VPA/1PIm)) using RAFT polymerization. The resulting P(VPA/1PIm)-block-polystyrene sample, produced in powder form, shows promise for PEM applications due to its robust thermal stability up to 300°C and its proton-conducting capabilities across a wide range of conditions (Suzuki et al., 2022). An imidazolium-based ionic liquid (BPMIm), featuring an extended 5-bromopentyl side chain, is synthesized and used as the quaternization agent for ether-free poly(p-terphenyl-co-N-methyl-piperidine) (PTP). By incorporating the flexible long-chain imidazolium group into the polymer backbone, the polymer's solubility in organic solvents is enhanced, while also introducing acid-base interaction sites. This structural enhancement facilitates phosphoric acid doping, crucial for achieving proton conductivity at temperatures above 100°C. The modified membrane exhibits anhydrous proton conductivity of 138 mS cm^{-1} at 180°C, attributed to increased phosphoric acid uptake. H_2-air fuel cells operating between 160–200°C demonstrate the viability of the PTP-PMIm membrane, achieving a peak power density of 456 mW cm^{-2} at ambient pressure (Jin et al., 2022). Norbornene-type PBI (NbPBI) (PILs) synthesized through *in-situ* free radical polymerization. Concurrently, the

PILs engage in a reaction with the norbornene monomer of NbPBI, resulting in the formation of a cross-linked network. Following the process of acid-base doping, the Br^- or Cl^- ions within the polymer ionic liquids are effectively substituted with $H_2PO_4^-$, thereby enhancing the transport of protons. The NbPBI-MPIm membrane exhibits a proton conductivity of 0.074 S cm^{-1} at 170°C, alongside phosphoric acid retention rates of 87.2% at 160°C and 73.8% at 80°C/40 RH%. The peak power density achieved was 385 mW cm^{-2} at a temperature of 160°C, accompanied by minimal voltage decay in the single cell following a 200-hour evaluation (Liu et al., 2021). Y. Feng and colleagues have introduced a promising method for creating anion-exchange membranes optimized for alkaline conditions. Their findings show that C2-substituted imidazolium PILs achieve remarkable ionic conductivity over 10^{-2} S cm^{-1} and maintain excellent stability for up to 60 hours in a 2 m KOH solution, without degradation (poly[DmVIm][Br]). However, when using PILs with imidazolium rings in fuel cells, careful preparation is crucial, as these rings are prone to ring-opening reactions under high-temperature, concentrated alkaline conditions, which can compromise membrane integrity (Lin et al., 2013; Liu et al., 2021). Yan and colleagues developed proton-conducting membranes by polymerizing microemulsions derived from PILs. The resulting PEMs demonstrated notable thermal stability and mechanical strength, achieving high conductivity around 1×10^{-1} S cm^{-1} at 160°C in anhydrous conditions. This performance is attributed to the well-connected PIL nanochannels within the material, which allow these PIL-based membranes to combine high proton conductivity with robust mechanical properties (Yan et al., 2009).

Despite the obvious advantages of PILs for proton exchange membrane in fuel cell (PEMFC) applications, difficulties like mechanical stability and ionic liquid leakage persist. Ongoing research is crucial to enhance these materials for practical application in fuel cells.

5.4 Dye Sensitized Solar Cells (DSSCs)

Dye-sensitized solar cells (DSSCs) are photo-electrochemical devices that transform light into energy, using light-sensitive semiconductor materials. The construction of these cells involves the integration of a photo-sensitized semiconductor anode with a meticulously designed electrolyte, leading to the formation of thin-film solar cells (Azaceta et al., 2010). Initial studies on incorporating poly(ionic liquids) (PILs) into solid-state DSSCs explored the distinctive properties of PILs containing iodide as a counter-anion. PILs address common issues linked to ionic liquid electrolytes, such as leakage and iodine evaporation, thereby improving the stability and longevity of DSSCs (Kawano et al., 2010). PILs, such as poly(3-decyl-1-vinyl imidazolium iodide) (PIL), when combined with polyindole and PMMA, create a porous membrane that significantly improves ionic conductivity and charge transfer, achieving a power conversion efficiency (PCE) of 5.96% (Thomas and Rajiv, 2020). PFII PILs were created through the use of 1-butylimidazolium iodide in poly(vinylidene fluoride-co-hexafluoropropylene) (PVDF-HFP). This produced a highly efficient electrolyte for quasi-solid-state (QSS) DSSCs by forming polymer membranes using a simple electrospinning technique. The findings indicate a superior open-circuit voltage (VOC), consistent short-circuit current density (JSC), an increased power conversion efficiency (η) of 9.26%, and better long-term stability (maintaining up to 97% of its initial η) over a duration of 1500 hours, compared to the cell employing a standard liquid electrolyte (8.63%) (Pang et al., 2018). Monomers of vinyl imidazolium-based ionic liquids with C4 and C8 alkyl tail lengths were synthesized through the quaternization of 1-vinylimidazole with the appropriate n-alkyl bromides. poly(1-vinyl-3-propylimidazolium). Iodide also demonstrated a comparable light-to-electricity conversion efficiency of 3.73% when exposed to 10 mW cm^{-2} (Chen et al., 2016). Poly(1-butyl-3-vinylimidazolium bis(trifluoromethane sulfonyl)-imide) (P[BVIM][TFSI]) is a photochemically stable PIL that Zhao et al. created in their quest for VOC-free IL/PILs electrolytes. It was then dissolved in an ionic liquid (RTIL) at room temperature. The P[BVIM][TFSI]-based electrolyte has a viscosity that is roughly 35 times higher than that of the liquid electrolyte, while the diffusion coefficient of I_3^- in the liquid electrolyte is 1.4 times greater than in the P[BVIM][TFSI]-based electrolyte (Zhao et al., 2011). At room temperature, poly(1-butyl-3-(1-vinylimidazolium-3-hexyl)-imidazolium bis(trifluoro methane-sulfonyl)imide) (P[BVIM][HIM][TFSI]) was more stable and conductive than mono-imidazolium-based electrolyte (P[BVIM][TFSI]). As a result of simulated air mass 1.5 sun spectrum illumination at 100 mW cm^2, the DSSCs were highly stable and had a power conversion efficiency of 5.92% (Chen et al., 2012). Despite the encouraging developments in PILs for DSSCs, issues persist in optimizing their performance across diverse environmental conditions and assuring cost-effectiveness for commercial applications. Additional study is required to resolve these difficulties and fully harness the potential of PILs in solar energy conversion.

Conclusion

This book chapter summarizes significant advancements in the synthesis of PILs and their development. The different mechanisms of polymerization reactions are elucidated by incorporating ionic liquids into their respective pathways. The brief explanation of production of poly ionic liquids also covered the development of PILs utilizing free radical polymerization techniques. There is a significant opportunity to expand the use of ionic liquid (IL) moieties by exploring new classes of monomers suitable for condensation polymerization. This process, which produces PIL materials, has the potential to greatly enhance the applications of ILs and broaden the range of products that can be derived from them. PILs exhibit considerable promise as advanced materials for energy storage, attributed to their distinctive ionic properties, structural flexibility, and adjustable chemistry. As electrolytes in batteries, supercapacitors, and various electrochemical devices, PILs enhance energy density, thermal stability, and safety. Although challenges persist, continued research and innovation are expected to reveal additional applications for PILs, positioning them as a fundamental component in the forthcoming generation of high-performance, sustainable energy storage technologies.

References

Appetecchi, G., Kim, G.-T., Montanino, M., Carewska, M., Marcilla, R., Mecerreyes, D. et al. 2010. Ternary polymer electrolytes containing pyrrolidinium-based polymeric ionic liquids for lithium batteries. Journal of Power Sources 195(11): 3668–3675.

Audic, N., Clavier, H., Mauduit, M. and Guillemin, J.-C. 2003. An ionic liquid-supported ruthenium carbene complex: a robust and recyclable catalyst for ring-closing olefin metathesis in ionic liquids. Journal of the American Chemical Society 125(31): 9248–9249.

Azaceta, E., Marcilla, R., Sanchez-Diaz, A., Palomares, E. and Mecerreyes, D. 2010. Synthesis and characterization of poly (1-vinyl-3-alkylimidazolium) iodide polymers for quasi-solid electrolytes in dye sensitized solar cells. Electrochimica Acta 56(1): 42–46.

Badri, M., Brunet, J.-J. and Perron, R. 1992. Ionic liquids as solvents for the regioselective O-alkylation of C/O ambident nucleophiles. Tetrahedron Letters 33(31): 4435–4438.

Bezdushna, E. and Ritter, H. 2009. Microwave and ionic liquids: inverse temperature dependence of viscosity in aqueous medium of grafted copolymers. Macromolecular Reaction Engineering 3(9): 516–521.

Bhavsar, R.S., Kumbharkar, S.C., Rewar, A.S. and Kharul, U.K. 2014. Polybenzimidazole based film forming polymeric ionic liquids: synthesis and effects of cation–anion variation on their physical properties. Polymer Chemistry 5(13): 4083–4096.

Carrasco, P.M., Ruiz de Luzuriaga, A., Constantinou, M., Georgopanos, P., Rangou, S., Avgeropoulos, A. et al. 2011. Influence of anion exchange in self-assembling of polymeric ionic liquid block copolymers. Macromolecules 44(12): 4936–4941.

Chang, H., Li, W., Liu, H., Hu, H., Liu, W., Jin, Y. et al. 2024. Molecular engineering of poly (ionic liquid)-based random copolymer electrolytes for enhanced performance of solid-state lithium batteries. Chemical Engineering Journal 481: 148602.

Chatel, G. and Macfarlane, D.R. 2014. Ionic liquids and ultrasound in combination: synergies and challenges. Chemical Society Reviews 43(23): 8132–8149.

Chauvin, Y., Mussmann, L. and Olivier, H. 1996. A novel class of versatile solvents for two-phase catalysis: hydrogenation, isomerization, and hydroformylation of alkenes catalyzed by rhodium complexes in liquid 1, 3-dialkylimidazolium salts. Angewandte Chemie International Edition in English 34(23-24): 2698–2700.

Chen, H., Choi, J.-H., Salas-de la Cruz, D., Winey, K.I. and Elabd, Y.A. 2009. Polymerized ionic liquids: the effect of random copolymer composition on ion conduction. Macromolecules 42(13): 4809–4816.

Chen, H. and Elabd, Y.A. 2009. Polymerized ionic liquids: solution properties and electrospinning. Macromolecules 42(9): 3368–3373.

Chen, S.-M., Wang, T.-L., Chang, P.-Y., Yang, C.-H. and Lee, Y.-C. 2016. Poly (ionic liquid) prepared by photopolymerization of ionic liquid monomers as quasi-solid-state electrolytes for dye-sensitized solar cells. Reactive and Functional Polymers 108: 103–112.

Chen, X., Zhao, J., Zhang, J., Qiu, L., Xu, D., Zhang, H. et al. 2012. Bis-imidazolium based poly (ionic liquid) electrolytes for quasi-solid-state dye-sensitized solar cells. Journal of Materials Chemistry 22(34): 18018–18024.

CJ, M. 2000. Smith PJ. Welton T. Chem. Commun 1249.

Csihony, S., Fischmeister, C., Bruneau, C., Horváth, I.T. and Dixneuf, P.H. 2002. First ring-opening metathesis polymerization in an ionic liquid. Efficient recycling of a catalyst generated from a cationic ruthenium allenylidene complex. New Journal of Chemistry 26(11): 1667–1670.

Das, A., Mei, B., Sokolov, A.P., Kumar, R. and Schweizer, K.S. 2024. Liquid state theory of the structure of model polymerized ionic liquids. The Journal of Chemical Physics 161(6).

Dong, K., Tang, W., Sun, Q., Chen, Z., Li, X., Zhai, D. et al. 2024. Rational design of bilayer heterogeneous poly (ionic liquid) s electrolytes for high-performance flexible supercapacitors with ultraslow self-discharge rate. Chemical Engineering Journal 153142.

Dou, J., Liu, Z., Mahmood, K. and Zhao, Y. 2012. Synthesis of poly (ethylene terephthalate) in benzyl imidazolium ionic liquids. Polymer International 61(9): 1470–1476.

Fan, F., Wang, W., Holt, A.P., Feng, H., Uhrig, D., Lu, X. et al. 2016. Effect of molecular weight on the ion transport mechanism in polymerized ionic liquids. Macromolecules 49(12): 4557–4570.

Fischer, T., Sethi, A., Welton, T. and Woolf, J. 1999. Diels-Alder reactions in room-temperature ionic liquids. Tetrahedron Letters 40(4): 793–796.

Green, M.D. and Long, T.E. 2009. Designing imidazole-based ionic liquids and ionic liquid monomers for emerging technologies. Polymer Reviews 49(4): 291–314.

Green, O., Grubjesic, S., Lee, S. and Firestone, M.A. 2009. The design of polymeric ionic liquids for the preparation of functional materials. Polymer Reviews 49(4): 339–360.

Gu, H., Yan, F. and Texter, J. 2016. Polymerized paired ions as polymeric ionic liquid–proton conductivity. Macromolecular Rapid Communications 37(14): 1218–1225.

Gunaratne, H.N., Langrick, C.R., Puga, A.V., Seddon, K.R. and Whiston, K. 2013. Production of polyetheretherketone in ionic liquid media. Green Chemistry 15(5): 1166–1172.

Gurkan, B.E., de la Fuente, J.C., Mindrup, E.M., Ficke, L.E., Goodrich, B.F., Price, E.A. et al. 2010. Equimolar CO_2 absorption by anion-functionalized ionic liquids. Journal of the American Chemical Society 132(7): 2116–2117.

Hirao, M., Ito, K. and Ohno, H. 2000. Preparation and polymerization of new organic molten salts; N-alkylimidazolium salt derivatives. Electrochimica Acta 45(8-9): 1291–1294.

Jin, Y., Che, X., Xu, Y., Dong, J., Pan, C., Aili, D. et al. 2022. An imidazolium type ionic liquid functionalized ether-free poly (terphenyl piperidinium) membrane for high temperature polymer electrolyte membrane fuel cell applications. Journal of The Electrochemical Society 169(2): 024504.

Kanetaka, Y., Yamazaki, S. and Kimura, K. 2016. Preparation of poly (ether ketone)s derived from 2, 5-furandicarboxylic acid by polymerization in ionic liquid. Macromolecules 49(4): 1252–1258.

Kaufmann, D.E., Nouroozian, M. and Henze, H. 1996. Molten salts as an efficient medium for palladium catalyzed CC coupling reactions. Synlett 1996(11): 1091–1092.

Kawano, R., Katakabe, T., Shimosawa, H., Nazeeruddin, M.K., Grätzel, M., Matsui, H. et al. 2010. Solid-state dye-sensitized solar cells using polymerized ionic liquid electrolyte with platinum-free counter electrode. Physical Chemistry Chemical Physics 12(8): 1916–1921.

Kim, T.Y., Lee, H.W., Stoller, M., Dreyer, D.R., Bielawski, C.W., Ruoff, R.S. et al. 2011. High-performance supercapacitors based on poly (ionic liquid)-modified graphene electrodes. ACS Nano 5(1): 436–442.

Korolovych, V.F., Erwin, A., Stryutsky, A., Lee, H., Heller, W.T., Shevchenko, V.V. et al. 2018. Thermally responsive hyperbranched poly (ionic liquid)s: assembly and phase transformations. Macromolecules 51(13): 4923–4937.

Kumbharkar, S.C., Bhavsar, R.S. and Kharul, U.K. 2014. Film forming polymeric ionic liquids (PILs) based on polybenzimidazoles for CO_2 separation. RSC Advances 4(9): 4500–4503.

Lei, Z., Dai, C., Zhu, J. and Chen, B. 2014. Extractive distillation with ionic liquids: a review. AIChE Journal 60(9): 3312–3329.

Li, X., Li, S., Zhang, Z., Huang, J., Yang, L. and Hirano, S.-i. 2016. High-performance polymeric ionic liquid–silica hybrid ionogel electrolytes for lithium metal batteries. Journal of Materials Chemistry A 4(36): 13822–13829.

Lin, B., Dong, H., Li, Y., Si, Z., Gu, F. and Yan, F. 2013. Alkaline stable C2-substituted imidazolium-based anion-exchange membranes. Chemistry of Materials 25(9): 1858–1867.

Lin, X., Liu, X., Tong, Y., Zhou, X., Li, J., Song, J. et al. 2024. A poly (ionic liquid)-based polymer binder for endurable lithium–sulfur batteries. Advanced Functional Materials 2406985.

Liu, F., Wang, S., Wang, D., Liu, G., Cui, Y., Liang, D. et al. 2021. Multifunctional poly (ionic liquid) s cross-linked polybenzimidazole membrane with excellent long-term stability for high temperature-proton exchange membranes fuel cells. Journal of Power Sources 494: 229732.

Liu, J., Xu, Y., Xu, F., Li, J., Chen, Y., Qiao, J. et al. 2023. Zwitterionic poly (ionic liquids)-based polymer electrolytes for Lithium-ion batteries applications. Ionics 29(6): 2249–2259.

Liu, X., Wen, Z., Wu, D., Wang, H., Yang, J. and Wang, Q. 2014. Tough BMIMCl-based ionogels exhibiting excellent and adjustable performance in high-temperature supercapacitors. Journal of Materials Chemistry A 2(30): 11569–11573.

Lu, F., Gao, X., Wu, A., Sun, N., Shi, L. and Zheng, L. 2017. Lithium-containing zwitterionic poly (ionic liquid)s as polymer electrolytes for lithium-ion batteries. The Journal of Physical Chemistry C 121(33): 17756–17763.

Lu, J., Yan, F. and Texter, J. 2009. Advanced applications of ionic liquids in polymer science. Progress in Polymer Science 34(5): 431–448.

Ma, X., Yu, J., Hu, Y., Texter, J. and Yan, F. 2023. Ionic liquid/poly (ionic liquid)-based electrolytes for lithium batteries. Industrial Chemistry & Materials 1(1): 39–59.

MacLean, B. and Singer, R. 1999. 1-Ethyl-3-methylimidazolium halogenoaluminate ionic liquids as solvents for Friedel–Crafts acylation reactions of ferrocene. Journal of the Chemical Society, Dalton Transactions (1): 63–66.

Mai, N.L. and Koo, Y.-M. 2016. Computer-aided design of ionic liquids for high cellulose dissolution. ACS Sustainable Chemistry & Engineering 4(2): 541–547.

Mallakpour, S. and Dinari, M. 2013. Straightforward and green method for the synthesis of nanostructure poly (amide-imide)s-containing benzimidazole and amino acid moieties by microwave irradiation. Polymer Bulletin 70(3): 1049–1064.

Marcilla, R., Blazquez, J.A., Fernandez, R., Grande, H., Pomposo, J.A. and Mecerreyes, D. 2005. Synthesis of novel polycations using the chemistry of ionic liquids. Macromolecular Chemistry and Physics 206(2): 299–304.

Marien, H., Bellings, L., Hermans, S. and Vankelecom, I.F. 2016. Sustainable process for the preparation of high-performance thin-film composite membranes using ionic liquids as the reaction medium. ChemSusChem 9(10): 1101–1111.

MatthewáReichert, W. 2002. Conventional free radical polymerization in room temperature ionic liquids: a green approach to commodity polymers with practical advantages. Chemical Communications (13): 1368–1369.

Mazur, A., Niesyto, K. and Neugebauer, D. 2022. Pharmaceutical functionalization of monomeric ionic liquid for the preparation of ionic graft polymer conjugates. International Journal of Molecular Sciences 23(23): 14731.

Mecerreyes, D. 2011. Polymeric ionic liquids: Broadening the properties and applications of polyelectrolytes. Progress in Polymer Science 36(12): 1629–1648.

Mu, X.-d., Meng, J.-q., Li, Z.-C. and Kou, Y. 2005. Rhodium nanoparticles stabilized by ionic copolymers in ionic liquids: long lifetime nanocluster catalysts for benzene hydrogenation. Journal of the American Chemical Society 127(27): 9694–9695.

Njoroge, I.G. 2017. Poly (Ionic Liquid) Thin Films via Surface-Initiated Ring-Opening Metathesis Polymerization. Vanderbilt University.

Njoroge, I., Matson, M.W. and Jennings, G.K. 2017. Dynamic anion-adaptive poly (ionic liquid) films via surface-initiated ring-opening metathesis polymerization. The Journal of Physical Chemistry C 121(37): 20323–20334.

Novakov, C., Kalinova, R., Veleva, S., Ublekov, F., Dimitrov, I. and Stoyanova, A. 2023. Flexible polymer-ionic liquid films for supercapacitor applications. Gels 9(4): 338.

Obadia, M., Colliat-Dangus, G., Debuigne, A., Serghei, A., Detrembleur, C. and Drockenmuller, E. 2015. Poly (vinyl ester 1, 2, 3-triazolium)s: a new member of the poly (ionic liquid)s family. Chemical Communications 51(16): 3332–3335.

Pang, H.-W., Yu, H.-F., Huang, Y.-J., Li, C.-T. and Ho, K.-C. 2018. Electrospun membranes of imidazole-grafted PVDF-HFP polymeric ionic liquids for highly efficient quasi-solid-state dye-sensitized solar cells. Journal of Materials Chemistry A 6(29): 14215–14223.

Peng, X., Liu, H., Yin, Q., Wu, J., Chen, P., Zhang, G. et al. 2016. A zwitterionic gel electrolyte for efficient solid-state supercapacitors. Nature Communications 7(1): 1–8.

Piotrowska, U., Sobczak, M. and Oledzka, E. 2017. Characterization of aliphatic polyesters synthesized via enzymatic ring-opening polymerization in ionic liquids. Molecules 22(6): 923.

Plechkova, N.V. and Seddon, K.R. 2008. Applications of ionic liquids in the chemical industry. Chemical Society Reviews 37(1): 123–150.

Pohako-Esko, K., Taaber, T., Saal, K., Lõhmus, R., Kink, I. and Mäeorg, U. 2013. New method for synthesis of methacrylate-type polymerizable ionic liquids. Synthetic Communications 43(21): 2846–2852.

Pont, A.-L., Marcilla, R., De Meatza, I., Grande, H. and Mecerreyes, D. 2009. Pyrrolidinium-based polymeric ionic liquids as mechanically and electrochemically stable polymer electrolytes. Journal of Power Sources 188(2): 558–563.

Qian, W., Texter, J. and Yan, F. 2017. Frontiers in poly (ionic liquid)s: syntheses and applications. Chemical Society Reviews 46(4): 1124–1159.

Rafiemanzelat, F. and Abdollahi, E. 2012. Rapid synthesis of new block copolyurethanes derived from L-leucine-PEG in ionic liquids under microwave irradiation. Journal of Applied Polymer Science 123(2): 1015–1023.

Rajagopal, R., Jarikote, D.V. and Srinivasan, K. 2002. Ultrasound promoted Suzuki cross-coupling reactions in ionic liquid at ambient conditions. Chemical Communications (6): 616–617.

Ray, A. and Saruhan, B. 2021. Application of ionic liquids for batteries and supercapacitors. Materials 14(11): 2942.

Ricks-Laskoski, H.L. and Snow, A.W. 2006. Synthesis and electric field actuation of an ionic liquid polymer. Journal of the American Chemical Society 128(38): 12402–12403.

Rojas, M.F., Bernard, F.L., Aquino, A., Borges, J., Dalla Vecchia, F., Menezes, S. et al. 2014. Poly (ionic liquid)s as efficient catalyst in transformation of CO_2 to cyclic carbonate. Journal of Molecular Catalysis A: Chemical 392: 83–88.

Ruckart, K.N., O'brien, R.A., Woodard, S.M., West, K.N. and Glover, T.G. 2015. Porous solids impregnated with task-specific ionic liquids as composite sorbents. The Journal of Physical Chemistry C 119(35): 20681–20697.

Salamone, J.C., Israel, S., Taylor, P. and Snider, B. 1973. Synthesis and homopolymerization studies of vinylimidazolium salts. Polymer 14(12): 639–644.

Sandoval, A.P., Feliu, J.M., Torresi, R.M. and Suárez-Herrera, M.F. 2014. Electrochemical properties of poly (3, 4-ethylenedioxythiophene) grown on Pt (111) in imidazolium ionic liquids. Rsc Advances 4(7): 3383–3391.

Sato, T., Morinaga, T., Marukane, S., Narutomi, T., Igarashi, T., Kawano, Y. et al. 2011. Novel solid-state polymer electrolyte of colloidal crystal decorated with ionic-liquid polymer brush. Advanced Materials 23(42): 4868.

Scovazzo, P., Havard, D., McShea, M., Mixon, S. and Morgan, D. 2009. Long-term, continuous mixed-gas dry fed CO_2/CH_4 and CO_2/N_2 separation performance and selectivities for room temperature ionic liquid membranes. Journal of Membrane Science 327(1-2): 41–48.

Shaplov, A.S., Marcilla, R. and Mecerreyes, D. 2015. Recent advances in innovative polymer electrolytes based on poly (ionic liquid)s. Electrochimica Acta 175: 18–34.

Stancik, C.M., Lavoie, A.R., Achurra, P.A., Waymouth, R.M. and Gast, A.P. 2004. A neutron scattering study of the structure and water partitioning of selectively deuterated copolymer micelles. Langmuir 20(21): 8975–8987.

Suga, T., Sakata, M., Aoki, K. and Nishide, H. 2014. Synthesis of pendant radical-and ion-containing block copolymers via ring-opening metathesis polymerization for organic resistive memory. ACS Macro Letters 3(8): 703–707.

Sun, Y., Wu, J. and Zhu, X. 2024. Poly (ionic liquids) membranes preparation and its application. Journal of Molecular Structure 137683.

Suzuki, Y., Nohara, T., Tabata, K., Yamakado, R., Shimada, R., Nakazaki, H. et al. 2022. Proton-conductive polymeric ionic liquids block copolymer of poly (vinylphosphonic acid)/1-propylimidazole-b-polystyrene for polymer electrolyte membrane fuel cells. Japanese Journal of Applied Physics 61(SD): SD1034.

Taghavi, M., Ghaemy, M., Nasab, S.M.A. and Hassanzadeh, M. 2013. Influence of ionic liquid on selective polycondensation of a new diamine-bisphenol: synthesis and properties of polyamides and their composites with modified nanosilica. Polymer 54(15): 3828–3840.

Taghavikish, M., Subianto, S., Gu, Y., Sun, X., Zhao, X. and Choudhury, N.R. 2018. A poly (ionic liquid) gel electrolyte for efficient all solid electrochemical double-layer capacitor. Scientific Reports 8(1): 10918.

Tang, H., Tang, J., Ding, S., Radosz, M. and Shen, Y. 2005a. Atom transfer radical polymerization of styrenic ionic liquid monomers and carbon dioxide absorption of the polymerized ionic liquids. Journal of Polymer Science Part A: Polymer Chemistry 43(7): 1432–1443.

Tang, J., Tang, H., Sun, W., Radosz, M. and Shen, Y. 2005b. Poly (ionic liquid) s as new materials for CO_2 absorption. Journal of Polymer Science Part A: Polymer Chemistry 43(22): 5477–5489.

Thomas, M. and Rajiv, S. 2020. Porous membrane of polyindole and polymeric ionic liquid incorporated PMMA for efficient quasi-solid state dye sensitized solar cell. Journal of Photochemistry and Photobiology A: Chemistry 394: 112464.

Tiruye, G.A., Munoz-Torrero, D., Palma, J., Anderson, M. and Marcilla, R. 2015. All-solid state supercapacitors operating at 3.5 V by using ionic liquid based polymer electrolytes. Journal of Power Sources 279: 472–480.

Vauthier, S., Alvarez-Tirado, M., Guzmán-González, G., Tomé, L., Cotte, S., Castro, L. et al. 2023. High-performance pyrrolidinium-based poly (ionic liquid) binders for Li-ion and Li-air batteries. Materials Today Chemistry 27: 101293.

Vijayakrishna, K., Jewrajka, S.K., Ruiz, A., Marcilla, R., Pomposo, J.A., Mecerreyes, D. et al. 2008. Synthesis by RAFT and ionic responsiveness of double hydrophilic block copolymers based on ionic liquid monomer units. Macromolecules 41(17): 6299–6308.

Vygodskii, Y.S., Lozinskaya, E.I. and Shaplov, A.S. 2002. Ionic liquids as novel reaction media for the synthesis of condensation polymers. Macromolecular Rapid Communications 23(12): 676–680.

Walden, P. 1914. Molecular weights and electrical conductivity of several fused salts. Bull. Acad. Imper. Sci.(St. Petersburg) 1800.

Wang, J. and Liu, Z. 2013. Ionic liquids as green reaction media for synthesis of poly (aryl ether ketone)s. Chinese Science Bulletin 58: 1262–1266.

Wang, J., Wu, Y. and Liu, Z.-p. 2016. A facile and highly efficient protocol for synthesis of poly (ether sulfone)s in ionic liquid/zwitterion. Chinese Journal of Polymer Science 34(8): 981–990.

Wang, J., Yang, Q., Liu, S., Tan, S., Wang, C. and Wu, Y. 2023. Protic poly (ionic liquid)/ionic liquid ionogel electrolyte sharing the same ionic species for supercapacitors. Industrial & Engineering Chemistry Research 62(51): 21977–21985.

Wang, T., Wang, W., Lyu, Y., Chen, X., Li, C., Zhang, Y. et al. 2017. Highly recyclable polymer supported ionic liquids as efficient heterogeneous catalysts for batch and flow conversion of CO_2 to cyclic carbonates. RSC Advances 7(5): 2836–2841.

Wei, X., Lin, T., Gao, J., Hu, Y., Zhang, Z., Peng, J. et al. 2024. Mechanically robust and highly conductive poly (ionic liquid)/polyacrylamide double-network hydrogel electrolytes for flexible symmetric supercapacitors with a wide operating voltage range. ACS Applied Materials & Interfaces 16(10): 12586–12598.

Wickramanayake, S., Hopkinson, D., Myers, C., Hong, L., Feng, J., Seol, Y. et al. 2014. Mechanically robust hollow fiber supported ionic liquid membranes for CO_2 separation applications. Journal of Membrane Science 470: 52–59.

Wiesenauer, E.F., Edwards, J.P., Scalfani, V.F., Bailey, T.S. and Gin, D.L. 2011. Synthesis and ordered phase separation of imidazolium-based alkyl–ionic diblock copolymers made via ROMP. Macromolecules 44(13): 5075–5078.

Wiesenauer, E.F., Nguyen, P.T., Newell, B.S., Bailey, T.S., Noble, R.D. and Gin, D.L. 2013. Imidazolium-containing, hydrophobic–ionic–hydrophilic ABC triblock copolymers: synthesis, ordered phase-separation, and supported membrane fabrication. Soft Matter 9(33): 7923–7927.

Xu, D., Guo, J. and Yan, F. 2018. Porous ionic polymers: Design, synthesis, and applications. Progress in Polymer Science 79: 121–143.

Yan, F., Yu, S., Zhang, X., Qiu, L., Chu, F., You, J. et al. 2009. Enhanced proton conduction in polymer electrolyte membranes as synthesized by polymerization of protic ionic liquid-based microemulsions. Chemistry of Materials 21(8): 1480–1484.

Yang, J., Sun, W., Lin, W. and Shen, Z. 2008. Synthesis and magnetic properties of comb-like copolymeric complexes based on thiazole ring and ionic liquid. Journal of Polymer Science Part A: Polymer Chemistry 46(15): 5123–5132.

Yao, Q. and Zhang, Y. 2003. Olefin metathesis in the ionic liquid 1-butyl-3-methylimidazolium hexafluorophosphate using a recyclable ru catalyst: remarkable effect of a designer ionic tag. Angewandte Chemie International Edition 42(29): 3395–3398.

Yin, K., Zhang, Z., Li, X., Yang, L., Tachibana, K. and Hirano, S.-i. 2015. Polymer electrolytes based on dicationic polymeric ionic liquids: application in lithium metal batteries. Journal of Materials Chemistry A 3(1): 170–178.

Yuan, J. and Antonietti, M. 2011. Poly (ionic liquid) s: Polymers expanding classical property profiles. Polymer 52(7): 1469–1482.

Yuan, J., Mecerreyes, D. and Antonietti, M. 2013. Poly (ionic liquid)s: An update. Progress in Polymer Science 38(7): 1009–1036.

Zhang, D.-Z., Ren, Y.-Y., Hu, Y., Li, L. and Yan, F. 2020. Ionic liquid/poly (ionic liquid)-based semi-solid state electrolytes for lithium-ion batteries. Chinese Journal of Polymer Science 38: 506–513.

Zhang, S., Féret, A., Lefebvre, H., Tessier, M. and Fradet, A. 2011. Poly (oxyalkylene) synthesis in brønsted acid ionic liquids. Chemical Communications 47(39): 11092–11094.

Zhao, H., Nathaniel, G.A. and Merenini, P.C. 2017. Enzymatic ring-opening polymerization (ROP) of lactides and lactone in ionic liquids and organic solvents: digging the controlling factors. RSC Advances 7(77): 48639–48648.

Zhao, J., Shen, X., Yan, F., Qiu, L., Lee, S. and Sun, B. 2011. Solvent-free ionic liquid/poly (ionic liquid) electrolytes for quasi-solid-state dye-sensitized solar cells. Journal of Materials Chemistry 21(20): 7326–7330.

Zhou, D., Liu, R., Zhang, J., Qi, X., He, Y.-B., Li, B. et al. 2017. *In situ* synthesis of hierarchical poly (ionic liquid)-based solid electrolytes for high-safety lithium-ion and sodium-ion batteries. Nano Energy 33: 45–54.

Zunita, M., Raizki, A., Aditya, R. and Wenten, I.G. 2022. Proton exchange polyionic liquid-based membrane fuel cell applications. Results in Engineering 16: 100653.

Chapter 6

Phenothiazine

Cutting-Edge Material for Next-Generation Energy Storage Solutions

Ayushi Bhatnagar, Rajendra Prasad Pakhariya and *Gangotri Pemawat**

1. Introduction

Energy storage technologies play a vital role in meeting the increasing demand for sustainable and dependable energy solutions. Organic compounds, in particular, have sparked interest because of their adaptable framework, configurable features, and environmental friendliness (Kebede et al., 2022). Phenothiazine (PTZ), a heterocyclic compound with a tricyclic structure, is a favorable scaffold for various applications. In pharmaceuticals, PTZ is crucial for antipsychotic medications like chlorpromazine and for use as anthelmintic in livestock (Otteny et al., 2020). In textile and ink industries, it serves as an important intermediate for dyes and pigments. Its conductive properties benefit organic electronics, including OLEDs and OPVs, while its applications in agriculture focus on crop protection and growth regulation (Gangadhar et al., 2021). The compound's redox-active nature enhances batteries and supercapacitors through efficient electron delocalization and reversible oxidation-reduction processes.

Department of Chemistry, University College of Science, Mohanlal Sukhadia University, Udaipur, Rajasthan, India 313001.

* Corresponding author: drgpemawat@mlsu.ac.in

DOI: 10.1201/9781003546672-6

PTZ-based electrodes exhibit good thermal and chemical stability, resulting in long cycle life and reduced replacement frequency (Rout et al., 2021). As an organic compound, PTZ is more sustainable than many traditional inorganic materials, and its synthesis is cost-effective. Tailoring PTZ derivatives can enhance properties like conductivity and solubility; this versatility allows them to be used in a wide range of applications, from portable electronics to large-scale grid storage (Wang et al., 2022). These derivatives can serve as cathode materials in LIBs and RFBs, and their fast electron transfer abilities also make them ideal for use in supercapacitors. Additionally, PTZ's peculiar structural and electronic properties make it an attractive material for BHJ solar cells. Its efficient charge transfer capabilities and tunable molecular structure allow optimization for Bulk heterojunction (BHJ) applications, resulting in strong light absorption and augmented power conversion efficiency when associated with compatible electron acceptors (Amin et al., 2024). Overall, PTZ-based polymers signify a major advancement in energy storage technology, offering a combination of elevated energy density, stability, and environmental compatibility, and will likely play an increasingly vital role in sustainable energy storage solutions (Chong et al., 2022).

2. PTZ-based Polymers

PTZ-based polymers are synthesized through various techniques that utilize the compound's redox-active properties. poly phenothiaizne (PPT) is produced via oxidative polymerization of PTZ monomers using oxidizing agents, forming a conjugated polymer (Wessling et al., 2024). PTZ-containing copolymers are developed by copolymerizing PTZ with other monomers, such as thiophene or aniline, through oxidative coupling or solution polymerization, resulting in polymers with tailored properties (Ravikiran and Chethan, 2022). PTZ-based polyimides are formed by polymerizing PTZ-containing diamines with dianhydrides in a high-temperature process, integrating PTZ into a thermally stable backbone. PTZ-functionalized polysiloxanes are synthesized by grafting PTZ onto polysiloxane backbones via chemical reactions like hydrosilylation, yielding flexible and durable polymers. These strategies improve the electrochemical capabilities of the polymers, making them well-adapted for numerous energy storage solutions (Lyu and Bielawski, 2022) (Fig. 1).

From 2015 to 2024, major strides have been made in adapting PTZ-based polymers for energy storage. Researchers have enhanced the stability, conductivity, and energy density of these materials through improved molecular design and polymerization techniques. These innovations have led to more efficient and durable polymers, positioning them as optimal contenders for advanced supercapacitors and batteries (Fig. 2).

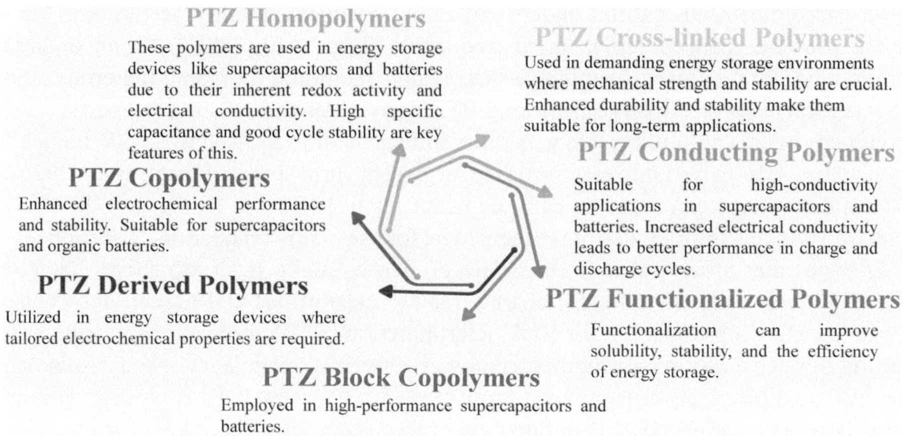

Fig. 1. Attributes and forms of PTZ-based polymers for energy storage.

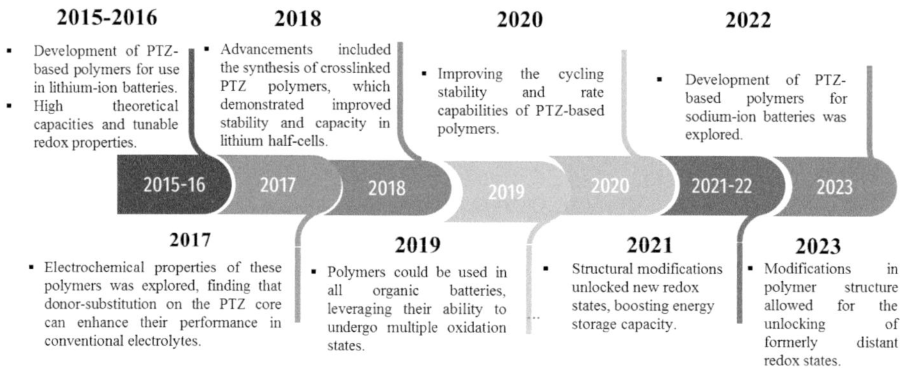

Fig. 2. Development timeline of PTZ-based energy storage devices.

2.1 Lithium-Ion Battery (LIB)

PTZ-based polymers have become strong prospects for LIBs thanks to their distinctive electrochemical properties, such as high specific capacity, exceptional redox reversibility, and strong cycling stability (Manthiram, 2020). These polymers, synthesized by incorporating PTZ units into the polymer backbone, exhibited enhanced performance over conventional materials like graphite and lithium cobalt oxide (Chen et al., 2021). The PTZ units facilitated efficient lithium-ion intercalation and deintercalation, while the polymer matrix provides a flexible, conductive network that supported ion transport and electron conduction. Additionally, PTZ-based polymers offered improved safety, being less prone to thermal runaway and dendrite formation, and can be derived from renewable resources, making them environmentally friendly (Edge et al., 2021). These advantages made PTZ-based polymers suitable for high-energy-density applications in portable electronics and wearable devices, positioning them as potential key players in the future of energy storage technology (Grey and Hall, 2020).

Zhang and his associates investigated the possibilities of two novel insoluble stiff hyper-crosslinked PTZ-functionalized polymers HPEPT and HPPT as organic cathode materials for LIBs. With discharge potentials of 3.65 V and 3.48 V, these PTZ-based p-type polymers exhibited reversible redox activity while maintaining the superior electrochemical characteristics of their parent monomers. The porous HPPT maintained strong electrical conductivity, enabling good redox kinetics and efficient usage of active sites (87% of the theoretical capacity) without the need for further changes. Its specific surface area was 452 m^2/g. The polymers' rigid structure also mitigated volume expansion and contraction, improving cycling stability. This development paved the way for new approaches in creating novel organic electrode materials (Zhang et al., 2020).

Bildirir et al. developed an economical approach for synthesizing a hyper-crosslinked PTZ polymer using a "knitting" polymerization technique based on the Friedel–Crafts mechanism. This polymer, characterized by a high density of PTZ redox units linked through allylic carbons, was tested as a cathode material for LIBs. It demonstrated a high capacity of 106 mAh/g at 0.5 C, a potential output of 3.6 V vs Li/Li^+, impressive rate capability (60 mAh/g at 15 C), and 79% capacity retention after 1000 cycles at 2 C. The efficient and straightforward synthesis method facilitated the scalable application of organic electrode materials while maintaining performance (Bildirir et al., 2023). Similarly, by expanding the conjugated structure of an active unit based on BPZ and its polymer p-DPBPZ, Li et al. were able to increase discharge voltage and stability. With discharge voltages between 3.20 and 3.45 V and 3.85 and 4.25 V (versus Li^+/Li), an initial specific capacity of 151 mAh/g, and energy and power densities of 537 W/kg and 1965 W/kg, respectively, the p-DPBPZ Li cells demonstrated remarkable performance. At 2 C, the average degradation rate of the extended conjugated structure was 0.0087% per cycle, resulting in high cycle stability due to prolonged delocalization of redox intermediates. Furthermore, p-DPBPZ showed promise as a sodium-ion battery active material. This study presented novel architectures and an effective development approach for organic electrode materials with good performance (Li et al., 2023).

2.2 Organic Radical Battery (ORB)

ORBs are an advanced innovation in energy storage, utilizing the distinct properties of organic radical compounds to deliver high power density, rapid charging, and eco-friendly benefits. ORBs employ stable organic radical molecules with unpaired electrons as active materials in both the cathode and the anode. Radicals like nitroxide undergo swift and reversible redox reactions, facilitating efficient charge and discharge cycles (Fig. 3). ORBs made of PTZ-based polymers represent an innovative leap in energy storage technology, combining the advantageous properties of organic radicals with the structural benefits of PTZ polymers. These batteries leverage the stable redox activity of PTZ units, which facilitate rapid and reversible electron transfer processes, resulting in high power density and fast

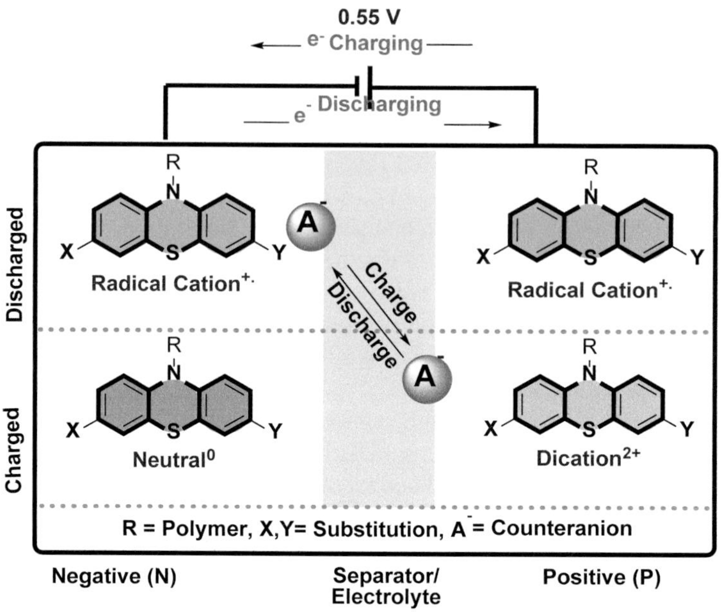

Fig. 3. Diagrammatic representation of ORB operation.

charging capabilities. The PTZ-based polymer matrix ensures robust mechanical flexibility and excellent cycling stability, reducing the degradation issues common in conventional battery materials. Additionally, these polymers can be synthesized from renewable resources, offering a more environmentally green and lasting alternative to traditional inorganic batteries. The unique combination of fast redox kinetics, high stability, and eco-friendly synthesis makes PTZ-based polymer ORBs a feasible solution for applications demanding reliable, efficient, and green energy storage.

Murugasen and his team developed a more sustainable rechargeable battery by synthesizing Lithium 4-(10H-phenothiazin-10-yl)benzoate (LiPHB), a monomeric p-type *N*-substituted PTZ salt. This compound functioned effectively as an anionic host in a dual-ion cell with lithium. When using a $LiClO_4$-based electrolyte, LiPHB exhibited a high operating potential of approximately 3.7 V vs. Li^+/Li and a reversible specific capacity of 86 mAh/g. It also demonstrated excellent cycling stability, maintaining its performance over 500 cycles at 0.2 C, positioning it as a promising option for advanced molecular organic batteries (Rajesh et al., 2020).

PTZ-based polymers have been synthesized; for example, PVMPT, was synthesized by Otteny and colleagues. These polymers were exceptional for their cycle stability and high redox potential of 3.5 V versus Li/Li^+. The main cause of this stability was the π-interactions between PTZ units, which allowed for a special charge/discharge process that involved the polymer's breakdown and redeposition. The effect of these π-interaction changes on battery performance was investigated. Pi-interactions were reduced by using dimeric reference compounds with bulky tolyl or mesityl substituents and the spatial separation between PTZ units was increased by adding maleimide groups into alternating copolymers. Electrochemical studies and UV/vis and EPR spectroscopy demonstrated that high cycle stability in PTZ-based poly(vinylene) derivative was largely dependent on the unique structure of PVMPT (Otteny et al., 2021).

2.3 Redox Flow Battery (RFB)

RFBs are a modern variant of rechargeable batteries where energy is maintained in liquid electrolytes located in external tanks, facilitating scalable energy storage solutions. These batteries were operated by circulating the electrolytes through a cell stack, where redox reactions occur at the electrodes to store and release energy (Weber et al., 2011). The two main components of RFBs are the electrochemical cell stack and the external storage tanks for the liquid electrolytes. RFBs made of PTZ-based polymers represent a significant advancement in energy storage technology, combining the modular scalability of RFBs with the electrochemical benefits of PTZ polymers (Wang et al., 2013). These polymers offered stable redox activity, high solubility, and excellent reversibility, leading to enhanced energy density and efficiency. The PTZ units in the polymer structure facilitated rapid electron transfer and stable redox cycling, which improved the overall performance and longevity of the battery (Pan and Wang, 2015). Additionally, PTZ-based polymers can be synthesized from renewable resources, making them an environmentally friendly alternative to traditional vanadium-based RFBs. The exceptional electrochemical properties of PTZ-based polymers, coupled with the scalability and safety benefits of redox flow systems, position them as a potential option for large-scale energy storage, particularly for renewable energy integration and grid stabilization (Noack et al., 2015). For non-aqueous RFBs, Yan et al. prepared hybrid catholytes that combined PTZ and DAC. These catholytes were created by employing a flexible C-N coupling technique to affix DAC substituent to the nitrogen of PTZ. A catholyte with stable two-electron cycling at 0.64 and 1.00 V vs. Fc/Fc^+ and good solubility ($\geq$ 0.45 M in $TBAPF_6$/MeCN) was identified by computational and structural investigations. This catholyte showed over 90% capacity retention after 266 hours of cycling with electron concentrations exceeding 0.5 M in high-energy-density RBFs. Another important development was the enhancement of a PTZ core by Attanayake and colleagues to get numerous electron transfers and good solubility. Two new PTZ derivatives were formed by them: PTZD-1 and 2 (Yan et al., 2021).

2.4 Organic Solar Cells (OSC)

OSCs, a revolutionary thin-film photovoltaic technology, have the potential to produce flexible, lightweight, and affordable devices through roll-to-roll production. In order to create a BHJ structure, OSCs often combine p-type and n-type semiconductors. The discovery of NFAs has dramatically raised PCE to values between 15–18%, which has sparked increasing interest in OSCs due to recent improvements (Fukuda et al., 2020). When designing active materials for OSCs, PTZ is useful, as it is a versatile dye that works well as an electron acceptor (n-type) or donor (p-type) in systems based on small molecules as well as polymers (Yao and Hou, 2022). In one study, Tang and colleagues synthesized two types of alternating copolymers, characterized as either regio-regular or regio-random, utilizing 10-alkylphenothiazine and 3-pentylthieno[3,2-b]thiophene as the electron donor components specifically for application in polymer solar cells. These copolymers yielded a maximum PCE of 0.2% (Boyer et al., 2008).

In a separate study, Lin and his research team developed a series of low-band-gap D–A copolymers that incorporated arylcyanovinyl-modified PTZ along with fluorene units. By fine-tuning the proportions and conjugation lengths of the D–A units, the team successfully enhanced intramolecular charge-transfer interactions, which broadened the absorption spectrum to the range of 400 to 800 nm in the visible spectrum. The resulting optical bandgaps of these materials ranged from 1.55 eV–2.10 eV. When applied in BHJ devices, these copolymers achieved a PCE of 0.51% (Li et al., 2008). Further advancements were reported by Chu et al., who presented an improved iteration of Lin's original polymer, which achieved a significant increase in efficiency, reaching up to 1.84% PCE (Huang et al., 2009). Additionally, Lin and his team explored a variety of low-band-gap polymers that included PTZ with various benzodiazole units, focusing on how structural variations impact their photovoltaic performance (Padhy et al., 2010).

2.5 Hole Transport Materials (HTMs)

HTMs are crucial for increasing the efficacy and performance of storage technologies, notably solid-state batteries and PSC. For successful charge transfer, holes, or positive charge carriers, must be able to travel from the active layer to the electrode which is made possible by HTMs (Yin et al., 2020). PTZ-based polymers have shown significant advances in electrochemical performance recently, which makes them very attractive for a variety of applications. The use of a PTZ unit as the main element of an HTM was introduced by Grisorio et al. They designed two HTMs with a PTZ core, adding a 4-methoxyphenyl group to improve heat stability, and used diarylamine and triarylamine as donor groups. According to their study, these HTMs might be significantly more affordable than spiro-OMeTAD while having comparable efficiency and stability. The influence of structural modifications was a significant finding: the sole difference between the two HTMs was the existence of phenyl spacers between the donor and core groups. One HTM only achieved 2.1% efficiency and demonstrated worse stability despite this tiny modification, highlighting the importance of the π-bridge between the PTZ core and the donor groups and the significant influence that small structural alterations impacting on performance (Grisorio et al., 2017). Following this, Liu et al. developed a similar HTM using a cost-effective, single-step process. They used triphenylamine moieties as hole-accepting groups and added octyl chains to the core's carbon atom to improve solubility and hydrophobicity. PSCs with this HTM achieved performance levels comparable to spiro-OMeTAD (Liu et al., 2017).

3. Recent Advances in PTZ-based Polymers

PTZ-based polymers have seen notable advancements in their electrochemical performance in recent times, which bodes well for their potential application in a range of energy storage technologies, including RFB, LIBs, and OSCs (Wang et al., 2023). Their latest performance attributes are outlined in Fig. 4.

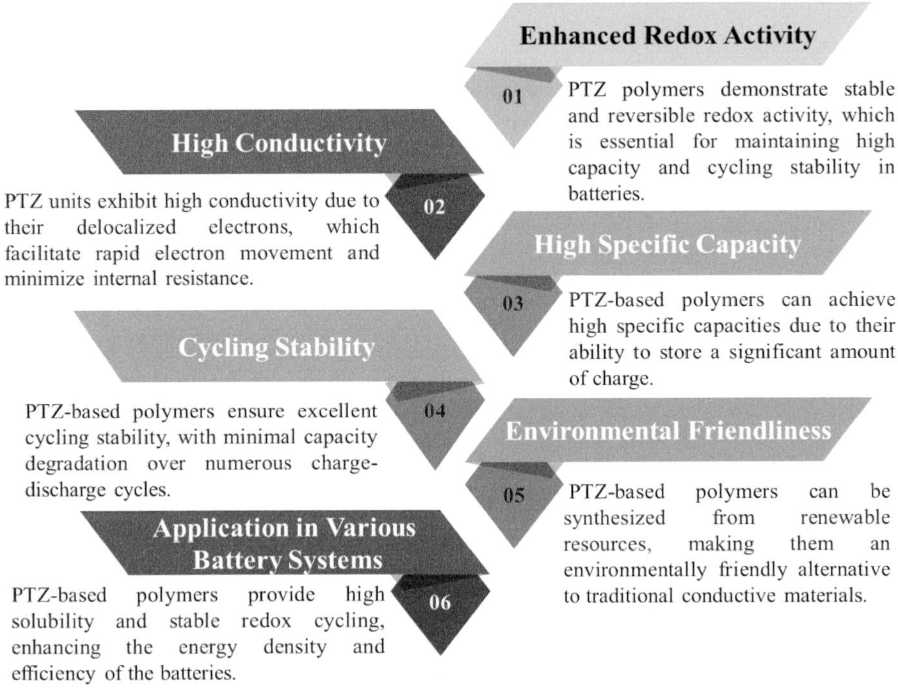

Fig. 4. Key performance traits of recent PTZ-based polymers for energy storage.

4. Conclusion

The unique electrochemical characteristics of PTZ-based polymers have shown them to have substantial potential in energy storage applications. These materials have qualities that are essential for efficient energy storage systems, including high stability, great conductivity, and changeable electrochemical properties. A variety of energy storage needs, including high capacity, prolonged cycle life, and quick charge-discharge performance, may be met by materials made with customized features due to the versatility of PTZ derivatives. It also provides room for additional performance optimization because these polymers' molecular structure may be changed. PTZ-based polymers should be crucial in the creation of sophisticated energy storage systems as research advances, propelling the development of high-performing and sustainable energy solutions.

Abbreviations:

PTZ	Phenothiazine
BHJ	Bulk heterojunction
OLED	Organic light emitting diode
OPV	Organic photovoltaics
PPT	Poly(PTZ)
HPEPT	Poly(1,3,5-tris((10-methyl-10*H*-phenothiazin-3-yl) ethynyl)benzene)
HPPT	Poly(1,3,5-tris(10-methyl-10*H*phenothiazin-3-yl)benzene)
LiPHB	Lithium 4-(10*H*-phenothiazin-10-yl)benzoate
BPZ	5,12-dihydrobenzo[*b*]phenazine
PVMPT	Poly(3-vinyl-*N*-methylphenothiaizne)
DAC	Diaminocyclopropenium
PTZD-1	N-ethyl-3,7-bis(2-(2-methoxyethoxy)ethoxy)phenothiazine
PTZD-2	3,7-dimethoxy-N-(2-(2methoxyethoxy)ethyl)phenothiazine.
spiro-OMeTAD	2,20,7,70-tetrakis[N,N-di(4-methoxyphenyl)amino]-9,90-spiro bifluorene
LBG	Low bandgap
D–A	Donor-acceptor
NFAs	Non-fullerene acceptors
PCE	Power conversion efficiencies
PSCs	Perovskite solar cells

References

Amin, M.F., Anwar, A., Gnida, P. and Jarząbek, B. 2024. Polymers containing phenothiazine, either as a dopant or as part of their structure, for dye-sensitized and bulk heterojunction solar cells. Polymers 16(16): 2309.

Attanayake, N.H., Kowalski, J.A., Greco, K.V., Casselman, M.D., Milshtein, J.D., Chapman, S.J. et al. 2019. Tailoring two-electron-donating phenothiazines to enable high-concentration redox electrolytes for use in nonaqueous redox flow batteries. Chemistry of Materials 31(12): 4353–4363.

Bildirir, H., Alvan, D., Patil, N., de la Peña O'Shea, V.A., Liras, M. and Marcilla, R. 2023. Hypercrosslinked phenothiazine polymer as a low-cost and durable organic cathode for rechargeable lithium batteries. ACS Applied Polymer Materials 6(17): 10092–10101.

Boyer, C., Liu, J., Wong, L., Tippett, M., Bulmus, V. and Davis, T.P. 2008. Stability and utility of pyridyl disulfide functionality in RAFT and conventional radical polymerizations. Journal of Polymer Science Part A: Polymer Chemistry 46(21): 7207–7224.

Chen, Y., Kang, Y., Zhao, Y., Wang, L., Liu, J., Li, Y. et al. 2021. A review of lithium-ion battery safety concerns: The issues, strategies, and testing standards. Journal of Energy Chemistry 59: 83–99.

Chong, K.F., Lee, S.P., Saufi, N.H.M. and Teo, E.Y.L. 2022. Phenothiazine-functionalized rGO for electrochemical capacitor. Current Science and Technology 2(2): 29–34.

Edge, J.S., O'Kane, S., Prosser, R., Kirkaldy, N.D., Patel, A.N., Hales, A. et al. 2021. Lithium ion battery degradation: what you need to know. Physical Chemistry Chemical Physics 23(14): 8200–8221.

Fukuda, K., Yu, K. and Someya, T. 2020. The future of flexible organic solar cells. Advanced Energy Materials 10(25): 2000765.

Gangadhar, P.S., Reddy, G., Prasanthkumar, S. and Giribabu, L. 2021. Phenothiazine functional materials for organic optoelectronic applications. Physical Chemistry Chemical Physics 23(28): 14969–14996.

Grey, C.P. and Hall, D.S. 2020. Prospects for lithium-ion batteries and beyond—a 2030 vision. Nature Communications 11(1): 1–4.

Grisorio, R., Roose, B., Colella, S., Listorti, A., Suranna, G.P. and Abate, A. 2017. Molecular tailoring of phenothiazine-based hole-transporting materials for high-performing perovskite solar cells. ACS Energy Letters 2(5): 1029–1034.

Huang, J.H., Li, K.C., Wei, H.Y., Chen, P.Y., Lin, L.Y., Kekuda, D. et al. 2009. Efficient bulk heterjunction solar cells based on a low-bandgap polyfluorene copolymers and fullerene derivatives. Organic Electronics 10(6): 1109–1115.

Kebede, A.A., Kalogiannis, T., Van Mierlo, J. and Berecibar, M. 2022. A comprehensive review of stationary energy storage devices for large scale renewable energy sources grid integration. Renewable and Sustainable Energy Reviews 159: 112213.

Li, K.C., Hsu, Y.C., Lin, J.T.S., Yang, C.C., Wei, K.H. and Lin, H.C. 2008. Novel narrow-band-gap conjugated copolymers containing phenothiazine-arylcyanovinyl units for organic photovoltaic cell applications. Journal of Polymer Science Part A: Polymer Chemistry 46(13): 4285–4304.

Li, M., Wang, R., Wu, T., Liu, Y., Chen, Y., He, W. et al. 2023. π-Extended Benzo [b] phenazine-based polymer cathode materials for high-voltage and stable organic batteries. ACS Applied Energy Materials 6(12): 6834–6841.

Liu, X., Tan, X., Chen, Q., Shan, H., Liu, C., Xu, J. et al. 2017. Facile synthesis of a dopant-free hole transporting material with a phenothiazine core for planar perovskite solar cells. RSC Advances 7(84): 53604–53610.

Lyu, J. and Bielawski, C.W. 2022. New classes of functionalized parylenes and poly (phenylene vinylene) s via coupling of dihaloxylyl diesters. Polymer Chemistry 13(5): 613–621.

Manthiram, A. 2020. A reflection on lithium-ion battery cathode chemistry. Nature Communications 11(1): 1550.

Noack, J., Roznyatovskaya, N., Herr, T. and Fischer, P. 2015. The chemistry of redox-flow batteries. Angewandte Chemie International Edition 54(34): 9776–9809.

Otteny, F., Desmaizieres, G. and Esser, B. 2020. Phenothiazine-based redox polymers for energy storage.

Otteny, F., Perner, V., Einholz, C., Desmaizieres, G., Schleicher, E., Kolek, M. et al. 2021. Bridging the gap between small molecular π-interactions and their effect on phenothiazine-based redox polymers in organic batteries. ACS Applied Energy Materials 4(8): 7622–7631.

Padhy, H., Huang, J.H., Sahu, D., Patra, D., Kekuda, D., Chu, C.W. et al. 2010. Synthesis and applications of low-bandgap conjugated polymers containing phenothiazine donor and various benzodiazole acceptors for polymer solar cells. Journal of Polymer Science Part A: Polymer Chemistry 48(21): 4823–4834.

Pan, F. and Wang, Q. 2015. Redox species of redox flow batteries: A review. Molecules 20(11): 20499–20517.

Rajesh, M., Dolhem, F., Davoisne, C. and Becuwe, M. 2020. Reversible anion insertion in molecular phenothiazine-based redox-active positive material for organic ion batteries. ChemSusChem 13(9): 2364–2370.

Ravikiran, Y. and Chethan, B. 2022. Humidity sensing studies on conducting polymers: Polyaniline and polypyrrole. Inorganic Chemistry Communications 145: 110019.

Rout, Y., Ekbote, A. and Misra, R. 2021. Recent development on the synthesis, properties and applications of luminescent oxidized phenothiazine derivatives. Journal of Materials Chemistry C 9(24): 7508–7531.

Wang, T., Chen, M., Sun, R. and Min, J. 2023. Recent research progress of all-polymer solar cells based on PSMA-type polymer acceptors. Chem 9(7): 1702–1767.

Wang, W., Luo, Q., Li, B., Wei, X., Li, L. and Yang, Z. 2013. Recent progress in redox flow battery research and development. Advanced Functional Materials 23(8): 970–986.

Wang, Y., Zhang, W., Yang, J., Gong, Y., Zhang, J., Fang, M. et al. 2022. The key role of molecular aggregation in rechargeable organic cathodes. Matter 5(12): 4467–4479.

Weber, A.Z., Mench, M.M., Meyers, J.P., Ross, P.N., Gostick, J.T. and Liu, Q. 2011. Redox flow batteries: a review. Journal of Applied Electrochemistry 41(10): 1137–1164.

Wessling, R., Delgado Andrés, R., Morhenn, I., Acker, P., Maftuhin, W., Walter, M. et al. 2024. Phenothiazine-based donor–acceptor polymers as multifunctional materials for charge storage and solar energy conversion. Macromolecular Rapid Communications 45(1): 2200699.

Yan, Y., Vogt, D.B., Vaid, T.P., Sigman, M.S. and Sanford, M.S. 2021. Development of high energy density diaminocyclopropenium-phenothiazine hybrid catholytes for non-aqueous redox flow batteries. Angewandte Chemie International Edition 60(52): 27039–27045.

Yao, H. and Hou, J. 2022. Recent advances in single-junction organic solar cells. Angewandte Chemie International Edition 61(37): e202209021.

Yin, X., Song, Z., Li, Z. and Tang, W. 2020. Toward ideal hole transport materials: a review on recent progress in dopant-free hole transport materials for fabricating efficient and stable perovskite solar cells. Energy & Environmental Science 13(11): 4057–4086.

Zhang, Y., Gao, P., Guo, X., Chen, H., Zhang, R., Du, Y. et al. 2020. Hypercrosslinked phenothiazine-based polymers as high redox potential organic cathode materials for lithium-ion batteries. RSC Advances 10(28): 16732–16736.

Chapter 7

Ferrocene-Based Redox Polymers for Medical Applications

Narendra Singh Chundawat

1. Introduction

Ferrocene, is an important example of organometallic chemistry, comprising two cyclopentadienyl anions ($C_5H_5^-$) sandwiching a central iron (Fe) atom. This sandwich structure was first discovered in 1951 by Kealy and Pauson. It indicates a significant advancement in the perception of metallocenes and their chemistry (Kealy et al., 1951). The extra stability and special electronic properties of ferrocene emerges from its aromatic cyclopentadienyl rings, which participate in a stable interaction with the iron center. Ferrocene displays remarkable redox behavior, with the Fe^{2+}/Fe^{3+} couple in solution and in solid-state, serving as a vital element for various applications (Bhatta et al., 2017).

Redox Polymers: Redox polymers are materials that can go through reversible oxidation and reduction reactions. Redox polymers are characterized by their ability to change their oxidation state in response to an external stimulus, which can be utilized for controlled and responsive functions (Bard et al., 2001). They generally contain redox-active groups that make easy electron transfer during these reactions. In case of polymer science, these materials are used in many applications such as sensors, batteries, and drug delivery systems.

Ferrocene-Based Redox Polymers: They are a branch of redox polymers where ferrocene units are incorporated into the polymer backbone. These polymers influence

Department of Chemistry, Bhupal Nobles' University, Udaipur 313002, Rajasthan, India.
Email: chundawat7@yahoo.co.in

DOI: 10.1201/9781003546672-7

the redox-active ferrocene units to create materials with special electrochemical properties. The synthesis of these polymers commonly involves polymerizing ferrocene-containing monomers or incorporating ferrocene derivatives into existing polymer structures.

2. Synthetic Methods

2.1 Introduction to Synthetic Methods

Ferrocene-based redox polymers have ferrocene units unified into the polymer structure. These polymers manifest special electrochemical properties because of ferrocene's redox activity making them appropriate for applications like drug delivery, diagnostics, and tissue engineering. Synthesis of such polymers includes many procedures, comprising polymerization of ferrocene monomers, post-polymerization modifications, and blending with other polymeric materials.

2.2 Ferrocene Monomers in Polymerization

The direct polymerization of ferrocene-containing monomers is a primary method for synthesizing ferrocene-based redox polymers. This approach can be broadly categorized into several polymerization techniques:

Free Radical Polymerization: For synthesizing ferrocene-based polymers free radical polymerization is a general procedure. This technique involves ferrocene derivatives as monomers and polymerization using free radical initiators. The general process consists of the following steps:

Synthesis of Monomer: Derivatives of ferrocene with polymerizable functional groups (e.g., vinyl, acrylate) are synthesized. General examples include ferrocene-based acrylates, methacrylates, or styrenes.

Polymerization: The monomers are polymerized in the presence of free radical initiators such as azobisisobutyronitrile (AIBN) or benzoyl peroxide. The reaction typically occurs in a solvent like toluene or dichloromethane (Scheme 1).

Example: The polymerization of ferrocene-based acrylates using AIBN as the initiator has been reported to yield polymers with high molecular weights and well-defined structures (Kim et al., 2023).

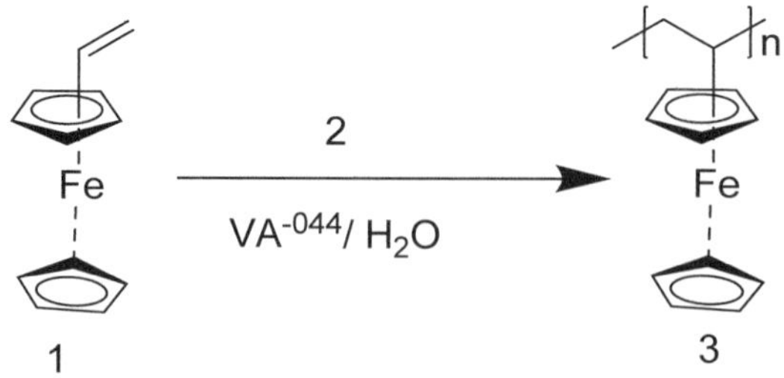

Scheme 1. Homopolymerization of vinylferrocene.

2.3 Coordination Polymerization

Coordination polymerization involves the use of transition metal catalysts to polymerize ferrocene derivatives. This method is particularly useful for creating polymers with well-defined architectures and controlled molecular weights.

Catalyst Selection: Catalysts such as metallocene complexes or other transition metal complexes are used. For instance, the use of ANSA-metallocene catalysts has been reported for the polymerization of ferrocene-based monomers (Sha et al., 2020).

Polymerization Conditions: Polymerization typically takes place in a non-aqueous solvent in an inert atmosphere to prevent oxidation (McGrath et al., 2019).

Example: To produce polymers with alternating structures coordination polymerization of ferrocene-based dienes using a zirconocene catalyst has been demonstrated (Wei and Diaconescu, 2019).

Ring-Opening Metathesis Polymerization (ROMP)

ROMP is another method used to polymerize ferrocene-containing monomers, particularly those with cycloolefin structures. This technique involves the use of metal-based catalysts to open and polymerize cyclic monomers (Scheme 2).

Designing Monomers: Ferrocene-based cycloolefins, like derivatives of cyclopentadienyl having pendant double bonds, are used (Straube et al., 2025).

Catalyst System: By ruthenium-based catalysts ROMP is commonly catalyzed, like those developed by Richard Schrock or Robert Grubbs (Ogba et al., 2020).

Example: Cyclooctenes based on Ferrocene have been polymerized by using ROMP to give polymers with controlled molecular weights and narrow polydispersities (Wei and Diaconescu, 2017).

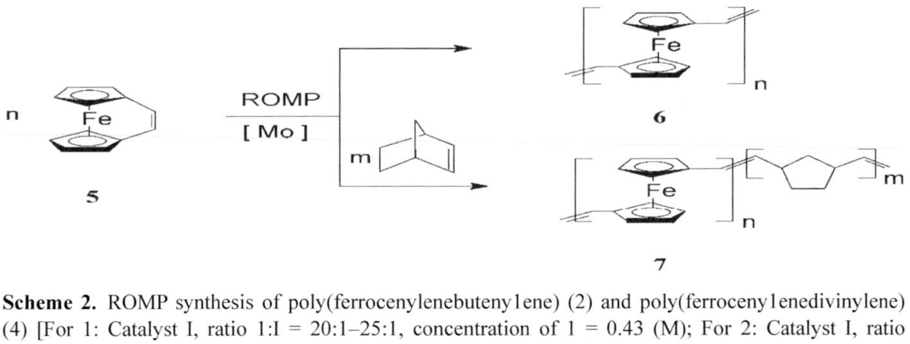

Scheme 2. ROMP synthesis of poly(ferrocenylenebutenylene) (2) and poly(ferrocenylenedivinylene) (4) [For 1: Catalyst I, ratio 1:I = 20:1–25:1, concentration of 1 = 0.43 (M); For 2: Catalyst I, ratio 2: I = 25:2–50:2, concentration of 2 = 0.42–2 (M)].

2.4 Post-Polymerization Functionalization

Post-polymerization functionalization involves modifying pre-existing polymers to incorporate ferrocene units. This approach is advantageous for introducing ferrocene into polymers that are otherwise difficult to polymerize directly.

Grafting Ferrocene onto Polymer Backbones: Grafting involves attaching ferrocene units to the side chains of a pre-existing polymer. This method can be achieved through:

Chemical Reactions: Reactions such as Friedel-Crafts alkylation or nucleophilic substitution can be used to attach ferrocene derivatives to polymer backbones (Amer et al., 2010).

Click Chemistry: Copper-catalyzed azide-alkyne cycloaddition (CuAAC) is used to attach ferrocene-based azides or alkynes to functionalized polymers (Ganesh, Sudhir et al., 2011). Example: Ferrocene-containing alkyl chains have been grafted onto polystyrene backbones using Friedel-Crafts alkylation, resulting in polymers with enhanced redox properties (Wang et al., 2019).

2.5 Crosslinking Ferrocene-Based Polymers

Crosslinking can enhance the mechanical properties and stability of ferrocene-based polymers. This method involves forming covalent bonds between polymer chains.

Crosslinking Agents: Agents such as bisphenols or multifunctional azides are used to create crosslinked networks. Conditions: Crosslinking typically occurs under heat or UV light, depending on the crosslinking agent used (Pietschnig, 2016). Example: Crosslinked ferrocene-based hydrogels have been synthesized for use in controlled drug delivery applications, demonstrating improved stability and release profiles (Wu et al., 2017).

Copolymerization and Blending

Copolymerization and blending methods involve combining ferrocene-containing monomers with other polymerizable units or polymers to achieve specific properties.

Copolymerization: Copolymerization involves polymerizing ferrocene-containing monomers with other monomers to create copolymers with tailored properties.

Monomer Selection: Ferrocene monomers are copolymerized with other functional monomers, such as those containing amines, esters, or other redox-active groups (Astruc, 2017).

Polymerization Techniques: Methods like free radical or coordination polymerization can be used for copolymerization (Madruga, 2002). Example: Copolymers of ferrocene-based methacrylates and acrylamide have been synthesized to enhance the responsiveness of the polymers for biomedical applications (Wu et al., 2017).

2.6 Blending with Other Polymers

Blending ferrocene-based polymers with other polymers or materials can create composites with improved properties.

Blending Methods: Physical blending or solvent casting techniques are used to combine ferrocene-based polymers with polymers like poly(lactic acid) (PLA) or poly(ethylene glycol) (PEG) (Yan et al., 2016).

Applications: These blends can be tailored for specific applications, such as drug delivery systems or tissue engineering scaffolds (Rauf et al., 2023). Example: Blends of ferrocene-based polymers with biodegradable polymers have been developed for controlled release applications, demonstrating effective drug encapsulation and release.

3. Medicinal Applications of Ferrocene-Based Redox Polymers

Ferrocene-based redox polymers are a dynamic area of research in medicinal chemistry due to their unique electrochemical properties, which offer significant potential for various medical applications. These polymers are characterized by their ferrocene units, which can undergo reversible redox reactions. This property enables a range of functionalities that are valuable in drug delivery, diagnostics, therapy, and tissue engineering. This comprehensive review explores the medicinal applications of ferrocene-based redox polymers, highlighting their synthesis, applications, and future prospects.

3.1 Introduction

Ferrocene, a metallocene compound consisting of two cyclopentadienyl rings bound to an iron atom, is known for its stable and reversible redox behavior. When incorporated into polymer structures, ferrocene endows the polymers with redox-active properties, making them useful for a variety of medicinal applications. The unique electrochemical characteristics of ferrocene-based polymers allow for responsive drug delivery systems, advanced diagnostic tools, innovative therapeutic agents, and functional tissue engineering scaffolds.

3.1.1 Redox Properties of Ferrocene

Ferrocene exhibits reversible redox behavior, undergoing oxidation to ferricinium (Fe^{3+}) and reduction back to ferrocene (Fe^{2+}). This redox activity is exploited in ferrocene-based polymers to create materials that can respond to external stimuli such as changes in pH, temperature, or electrical signals. This responsiveness is crucial for developing advanced medical applications where controlled behavior and targeting are necessary.

3.2 Types of Ferrocene-Based Polymers

Ferrocene-based polymers can be classified based on their structure and functionality:

Homopolymers: Composed of a single type of ferrocene derivative.

Copolymers: Incorporate ferrocene units along with other monomers to modify properties.

Grafted Polymers: Ferrocene units attached to a pre-existing polymer backbone.

Crosslinked Polymers: Forms networks with enhanced mechanical properties and stability.

Each type of polymer has distinct properties and applications, which will be detailed in the following sections.

3.3 Drug Delivery Systems

3.3.1 Controlled Release Systems

Ferrocene-based redox polymers offer a robust platform for controlled drug release. These polymers can be engineered to release drugs in response to redox stimuli, which can be triggered by external electric fields or internal biological conditions.

Mechanism: The release mechanism typically involves embedding the drug within a ferrocene-containing polymer matrix. Upon applying a redox stimulus, the polymer's structure changes, which can result in the controlled release of the drug (Huo et al., 2014). This mechanism allows for precise control over the timing and rate of drug release.

Ferrocene-Loaded Hydrogels: These hydrogels can release encapsulated drugs when subjected to an oxidative or reductive environment. For instance, a ferrocene-based hydrogel was designed to release anticancer drugs in response to oxidative stress (Ornelas and Astruc, 2023).

Ferrocene-based Micelles: Self-assembled micelles incorporating ferrocene units can be used for controlled drug delivery. These micelles can undergo structural changes in response to redox stimuli, releasing their payload in a controlled manner (Xu et al., 2017).

3.3.2 Targeted Drug Delivery

Targeted drug delivery systems using ferrocene-based polymers involve functionalizing the polymer to recognize and bind specific biological targets. The redox activity of ferrocene can be employed to release drugs at targeted sites in the body.

Mechanism: Functional groups or targeting ligands are conjugated to the ferrocene-based polymer, allowing for selective binding to specific cells or tissues. The redox properties can then trigger drug release precisely at the target site (Ornelas and Astruc, 2023).

Cancer Therapy: Ferrocene-based polymers functionalized with targeting ligands (e.g., antibodies) have been developed to deliver chemotherapeutic agents specifically to cancer cells, minimizing side effects and improving therapeutic efficacy (Wu et al., 2025).

Inflammatory Diseases: Targeted delivery systems for inflammatory diseases have been designed using ferrocene-based polymers, enabling localized treatment and reducing systemic drug exposure (Sharma and Rathore, 2023).

3.4 Diagnostic Applications

3.4.1 Biosensors

Ferrocene-based polymers are utilized in the development of biosensors because of their electrochemical properties. These sensors can detect a wide range of biological substances by measuring changes in the redox state of ferrocene units.

Mechanism: In a biosensor, the ferrocene-based polymer interacts with a target biomolecule, leading to a measurable change in the redox signal. This signal change can be translated into a concentration measurement or detection of the biomolecule (Saleem et al., 2015).

Glucose Sensors: Ferrocene-based polymers have been used to create highly sensitive glucose sensors. These sensors work by detecting changes in the redox state of ferrocene in the presence of glucose, providing real-time glucose monitoring (Merchant et al., 2009).

DNA Sensors: Ferrocene-based biosensors for DNA detection utilize the redox properties of ferrocene to detect specific DNA sequences, offering a sensitive and selective method for genetic analysis (Roy et al., 2022).

3.4.2 Imaging and Detection

Ferrocene-based polymers are also explored for imaging applications. Their electrochemical properties can enhance contrast or signal in various imaging techniques.

Mechanism: By incorporating ferrocene units into imaging agents or contrast materials, the redox properties of these units can improve imaging contrast or signal. This is particularly useful in techniques like magnetic resonance imaging (MRI) and electrochemical impedance spectroscopy (Zhou et al., 2023).

MRI Contrast Agents: Ferrocene-based nanoparticles have been developed to enhance MRI imaging. These nanoparticles provide improved contrast due to the highly flexible nature of ferrocene (Takahashi and Anzai, 2013).

Electrochemical Imaging: Ferrocene-based polymers are used in electrochemical imaging techniques to provide high-resolution images of biological samples (Sun et al., 2014).

3.5 Therapeutic Applications

3.5.1 Anticancer Agents

Ferrocene-based polymers exhibit significant potential as anticancer agents. The redox properties of these polymers can be harnessed to develop novel therapeutic agents with enhanced anticancer activity.

Mechanism: Ferrocene-based compounds can induce oxidative stress in cancer cells, leading to apoptosis or inhibition of tumor growth. Additionally, these polymers can be used to deliver chemotherapeutic agents more effectively (Lin et al., 2023).

Ferrocene-Based Polymers with Cytotoxicity: Several studies have reported the cytotoxic effects of ferrocene-based polymers against various cancer cell lines. These polymers can disrupt cellular processes and induce cancer cell death (Larik et al., 2017).

Targeted Anticancer Therapy: Ferrocene-based polymers functionalized with targeting ligands have been used to deliver anticancer drugs specifically to tumor cells, enhancing the therapeutic efficacy and reducing side effects (Sharma et al., 2025).

3.5.2 Antimicrobial Agents

Ferrocene-based polymers also show promise as antimicrobial agents. Their redox properties can be used to develop materials with antimicrobial activity, effective against a broad range of pathogens.

Mechanism: The antimicrobial activity of ferrocene-based polymers is often attributed to their ability to generate reactive oxygen species or disrupt microbial cell membranes (Costa et al., 2020).

Ferrocene-containing coatings: Antimicrobial coatings incorporating ferrocene units have been developed for use in medical devices and surfaces. These coatings provide protection against bacterial and fungal contamination (Woo et al., 2021).

Wound Dressings: Ferrocene-based wound dressings with antimicrobial properties have been created to prevent infections and promote healing in chronic wounds (Li et al., 2024).

3.6 Tissue Engineering and Regenerative Medicine

3.6.1 Scaffolds for Tissue Regeneration

Ferrocene-based polymers are used to create scaffolds for tissue engineering. These scaffolds provide a supportive matrix for cell growth and tissue regeneration, with the added advantage of redox responsiveness.

Mechanism: The mechanical properties and biocompatibility of ferrocene-based polymers can be tuned to create scaffolds that mimic the extracellular matrix. The redox properties can also be used to stimulate cell growth or differentiation.

Ferrocene-Based Hydrogels: Hydrogels incorporating ferrocene units have been used as scaffolds for tissue engineering. These hydrogels can support the growth of various cell types and promote tissue regeneration.

Ferrocene-Modified Scaffolds: Scaffolds modified with ferrocene units have been developed for use in bone and cartilage tissue engineering. These materials enhance cell adhesion and proliferation, supporting tissue repair.

3.6.2 Injectable Hydrogels

Injectable ferrocene-based hydrogels offer a minimally invasive approach to tissue repair. These hydrogels can be injected into damaged tissues and then undergo gelation to form a supportive matrix.

Mechanism: Injectable hydrogels can gel in response to physiological conditions such as pH or temperature changes. The redox properties of ferrocene can be used to control the release of growth factors or other bioactive molecules within the hydrogel.

Bone Repair: Injectable ferrocene-based hydrogels have been used for bone repair, providing a scaffold that supports bone regeneration and integration with the surrounding tissue.

Cartilage Repair: Ferrocene-based hydrogels have also been developed for cartilage repair, offering a flexible and responsive material for regenerating cartilage tissue. Drug therapeutic applications for ferrocene and ferrocene derivatives have been tabulated in Table 1.

Table 1. Ferrocene and Ferrocene derivative for medical application.

Ferrocene/Ferrocene Derivative	Therapeutic Application as drugs	Reference
Ferroquine	An antimalarial agent effective against chloroquine-resistant strains of Plasmodium falciparum.	(Debarshi Kar Mahapatra et al., 2015)
Ferrocifen	A ferrocene-modified derivative of tamoxifen, exhibiting significant anticancer activity, particularly against breast cancer cell lines.	(Ca´tia Ornelas, 2011)
Ferrocene-based Anticancer Agents	Anticancer properties, showing promising results in inhibiting tumor growth.	(Amandeep Singh et al., 2019)
Ferrocene-based Antiparasitic Compounds	Efficacy against various parasitic infections, including malaria and trypanosomiasis.	(Amandeep Singh et al., 2019)
Ferrocene-containing Anti-inflammatory Agents	Potential as anti-inflammatory agents by modulating oxidative stress pathways.	(Amandeep Singh et al., 2019)

Table 1 contd. ...

...Table 1 contd.

Ferrocene/Ferrocene Derivative	**Therapeutic Application as drugs**	**Reference**
Ferrocene-linked Antioxidants	Compounds where ferrocene is conjugated with antioxidant moieties, enhancing their ability to scavenge free radicals.	(Ca´tia Ornelas, 2011)
Ferrocene-based HIV-1 Protease Inhibitors	Potential to inhibit HIV-1 protease, a key enzyme in the life cycle of HIV.	(Ca´tia Ornelas, 2011)
Ferrocene-derived Antitubercular Agents	Activity against *Mycobacterium tuberculosis*, the causative agent of tuberculosis.	(Ca´tia Ornelas, 2011)
Ferrocene-based Antiviral Agents	Potential antiviral agents against various viruses.	(Ca´tia Ornelas, 2011)
Ferrocene-containing Antifungal Agents	Capacity to combat fungal infections.	(Ca´tia Ornelas, 2011)
Ferrocene-based Enzyme Inhibitors	Potential to inhibit specific enzymes involved in disease pathways.	(Ca´tia Ornelas, 2011)
Ferrocene-based Radiopharmaceuticals	Potential applications in diagnostic imaging and therapy	(Ca´tia Ornelas, 2011)
Ferrocene-containing Antimalarial Agents	Beyond ferroquine, other ferrocene derivatives have been investigated for antimalarial activity.	(Ca´tia Ornelas, 2011)
Ferrocene-derived Neuroprotective Agents:	Potential to protect neurons from oxidative stress	(Ca´tia Ornelas, 2011)
Ferrocenyl Tamoxifen (Ferrocifen)	Efficacy in breast cancer therapy	(Ca´tia Ornelas, 2011)
Ferrocenyl Quinones	Capacity to induce apoptosis in cancer cells	(Kovalskaa, 2014)
Ferrocenyl Chalcones	Capacity for cytotoxicity against tumors	(Tsang-Lin Hwang, 2021)
Ferrocenyl Curcuminoids	Antioxidant, anti-inflammatory properties	(Esmaeil Niknam, 2020)
Ferrocenyl Anthraquinones	DNA intercalation, anticancer properties	(Kovalskaa, 2014)
Ferrocenyl Betulinic Acid	Anti-HIV and anticancer properties	(Ayumu Niida, 2017)
Ferrocenyl Indoles	Capacity to Target DNA synthesis	(Coralie C. Quadr, 2018)
Ferrocenyl Thioureas	Capacity to block cancer cell proliferation	(Mikhail Krasavin, 2016)
Ferroquine (SSR97193)	Activity against Plasmodium falciparum	(Debarshi Kar Mahapatra, 2015)
Ferrocenyl Triazoles	Antifungal (Candida) properties	(Yue Yuan et al., 2019)

Table 1 contd. ...

...*Table 1 contd.*

Ferrocene/Ferrocene Derivative	**Therapeutic Application as drugs**	**Reference**
Ferrocenyl Fluoroquinolones	Antibacterial (E. coli, S. aureus) properties	(Soraya Alnabulsi et al., 2018)
Ferrocenyl Benzimidazoles	Broad-spectrum antimicrobial properties	(Kirill Nikitin et al., 2019)
Ferrocenyl Sulfonamides	Antibacterial resistance inhibition	(Christiana M. Adeyemi et al., 2020)
Ferrocenyl NSAID Hybrids	Cyclooxygenase-2 (COX-2) inhibition	(Amandeep Singh et al., 2019)
Ferrocenyl Vanillin	Biofilm inhibition	(Muhammad Usman et al., 2020)
Ferrocenyl Thiosemicarbazones	SARS-CoV-2 inhibition	(Rifat Jawarial et al., 2021)
Ferrocenyl Thioureas	Hepatitis virus inhibition	(Mikhail Krasavin et al., 2016)
Ferrocenyl Methionine	Potential in Alzheimer's treatment	(Meng Li et al., 2017)
Ferrocenyl Peptides	Antimicrobial properties	(Xuejiao Yang et al., 2017)
Ferrocene-quinoline conjugates	Anti-microbial activity against Gram-positive and Gram negative bacteria, as well as human pathogen yeast C	(Amandeep Singh et al., 2019)
Ferrocene- pyrazoline derivatives	Activity against human, animal and plant diseases, mycotoxin producers, and food spoilage agents.	(Amandeep Singh et al., 2019)
Ferrocene-1H-1,2,3-triazoles	Antiinflammatory activities.	(Amandeep Singh et al., 2019)
Isatin-ferrocenylchalcone chalcones	Antimicrobial evaluation against the human mucosal pathogen Trichomonas vaginalis	(Amandeep Singh et al., 2019)
Ferrocenyl hydrazones	Anti-tubercular activities	(Amandeep Singh et al., 2019)
Ferrocenyl thiosemicarbazones	Anti-tubercular activities.	(Amandeep Singh et al., 2019)
Ferrocenyl-benzimidazole	Anti-malarial properties	(Amandeep Singh et al., 2019)
Ferrocene nitrophenyl conjugates	Efficacy as panel of cancer cell line using cisplatin as reference drug	(Amandeep Singh et al., 2019)

Table 1 contd. ...

...Table 1 contd.

Ferrocene/Ferrocene Derivative	**Therapeutic Application as drugs**	**Reference**
Ferrocenyl analogues of clotrimazole	Antitumor profiles against breast cancer cell MCF-7 and colon cancer cell line HT29.	(Amandeep Singh et al., 2019)
Ferrocenequinonemethide-thiol adducts	Exhibited antiproliferative activities against hormone-resistant cancer cells with effects being comparable to that of ferrociphenols	(Amandeep Singh et al., 2019)
Aceylated-hydroxypropyl-ferrociphenols	Antiproliferative activity against MDA-MB-231 breast cancer cell line and have shown good activity with IC50s in micro to sub-micromolar.	(Amandeep Singh et al., 2019)
4-aminoquinoline-ferrocenyl-chalcone conjugates	Antiplasmodial evaluation	(Amandeep Singh et al., 2019)
β-lactamferrocenylchalcones	Antimalarial profiles against the CQ-S 3D7 and CQ-R W2 strain of P. falciparum	(Amandeep Singh et al., 2019)

4. Challenges and Future Directions

Despite the promising applications of ferrocene-based redox polymers, several challenges need to be addressed:

4.1 Biocompatibility

Ensuring the biocompatibility of ferrocene-based polymers is crucial for their medical use. This includes evaluating their potential to induce adverse reactions in the body and ensuring they do not interfere with normal physiological processes.

4.2 Stability

The stability of ferrocene-based polymers in physiological conditions must be carefully assessed. Polymers should maintain their redox activity and mechanical properties over time to ensure their effectiveness and safety (Hu et al., 2020).

4.3 Scalability

Developing scalable synthesis methods for ferrocene-based polymers is necessary for their commercialization and widespread use. Research should focus on optimizing synthesis routes and reducing production costs (Saleem et al., 2015).

4.4 Regulatory Considerations

Compliance with regulatory requirements is essential for the clinical application of ferrocene-based polymers. This includes ensuring that the materials meet safety and efficacy standards set by regulatory agencies (Patra and Gasser, 2017).

5. Conclusion

Ferrocene-based redox polymers represent a significant advancement in the field of medical applications, blending the unique properties of ferrocene with the versatile nature of polymer science. This intersection of chemistry and materials science opens up promising avenues for innovation in healthcare technologies.

First and foremost, the intrinsic redox properties of ferrocene, coupled with its stability and biocompatibility, offer a robust platform for the development of advanced biomedical devices. The ability to modulate redox states with precision facilitates the creation of electroactive polymers that can be finely tuned for specific medical applications. These materials are not only capable of delivering drugs in a controlled manner but also of providing real-time monitoring and responsive therapy through electrochemical signals.

In drug delivery systems, ferrocene-based polymers enhance the functionality and efficiency of therapeutic agents. The incorporation of redox-active groups into polymer matrices allows for the development of systems that can respond dynamically to physiological conditions, thereby improving drug release profiles and reducing side effects. This capability is particularly advantageous in targeted therapies where localized and controlled drug delivery is crucial.

Moreover, the use of ferrocene-based redox polymers in biosensing applications demonstrates their potential to revolutionize diagnostic technologies. Their ability to interact with biological molecules through electrochemical reactions provides a sensitive and specific approach to detecting biomarkers and other analytes. This could lead to the development of more accurate and rapid diagnostic tools, which are essential for early disease detection and personalized medicine.

In addition, the biocompatibility and ease of functioning of ferrocene-based polymers make them suitable candidates for use in implantable devices and tissue engineering. These materials can be engineered to integrate seamlessly with biological tissues, promoting better integration and reducing the risk of adverse reactions. Furthermore, their electroactivity can be harnessed to stimulate cellular responses, supporting tissue regeneration and repair.

The versatility of ferrocene-based redox polymers extends to their use in the development of novel therapeutic strategies. Their ability to undergo reversible redox reactions can be exploited to design smart therapeutic systems that adapt to changing physiological conditions, offering new possibilities for the treatment of chronic diseases and complex health conditions.

Despite the considerable promise of ferrocene-based redox polymers, there are challenges that need to be addressed. The scalability of production, long-term stability, and the potential for bioaccumulation are important factors that must be carefully considered. Continued research and development are necessary to optimize these materials for clinical use, ensuring their safety and efficacy in real-world medical applications.

In conclusion, ferrocene-based redox polymers represent a transformative technology in medical science. Their unique properties offer numerous advantages across a range of applications, from drug delivery and biosensing to tissue engineering and beyond. As research progresses and these materials become more refined, they hold the potential to significantly impact patient care and treatment outcomes. The future of ferrocene-based redox polymers is promising, and their continued exploration will undoubtedly yield innovative solutions that enhance the quality of medical care and improve the lives of patients worldwide.

References

Adeyemi, C.M., Conibear, A.C.., Muttorwa, M.K., Nokalipa, I.C., Isaacs, M., Mnkandhla, D. et al. 2020. Synthesis and anti-parasitic activity of achiral N-benzylated phosphoramidic acid derivatives. Bioorganic Chemistry 101: 103947.

Alnabulsi, S., Hussein, B., Santina, E., Alsalahat, I., Kadirvel, M., Magwaza, R. et al. 2018. Evaluation of analogues of furan-amidines as inhibitors of NQO2. Bioorganic & Medicinal Chemistry Letters 28(8): 1292–1297.

Amer, W.A., Abid, M.A.,Wang, L., Ma, L. and Yu, H. 2010. Recent progress in the synthesis and applications of some ferrocene derivatives and ferrocene-based polymers. Journal of Inorganic and Organometallic Polymers and Materials 20: 605–615.

Anusionwu, C.G. 2021. Synthesis, Characterisation and *in Vitro* Evaluation of Hybrid Compounds Containing Ferrocene and Nitrogen-Containing Bisphosphonates, University of Johannesburg (South Africa).

Astruc, D. 2017. Why is ferrocene so exceptional? European Journal of Inorganic Chemistry 2017(1): 6–29.

Bard, A.J., Faulkner, L.R. and White, H.S. 2022. Electrochemical Methods: Fundamentals and Applications. John Wiley & Sons.

Bhatta, S.R., Bheemireddy, V., Vijaykumar, G., Debnath, S. and Thakur, A. 2017. An efficient molecular tool with ferrocene backbone: discriminating Fe^{3+} from Fe^{2+} in aqueous media. Organometallics 36(11): 2141–2152.

Butler, I. 1994. 2.0 Monocyclopentadienyls. Organometallic Chemistry: 23: 376.

Chauhan, N.P.S. and Chundawat, N.S. 2019. Inorganic and Organometallic Polymers, Walter de Gruyter GmbH & Co KG.

Costa, N.C., Piccoli, J.P., Santos-Filho, N.A., Clementino, L.C., Almeida, A.M., De Annunzio, S.R. et al. 2020. Antimicrobial activity of RP-1 peptide conjugate with ferrocene group. Plos One 15(3): e0228740.

Fujimura, K., Ouchi, M. and Sawamoto, M. 2015. Ferrocene cocatalysis for iron-catalyzed living radical polymerization: Active, robust, and sustainable system under concerted catalysis by two iron complexes. Macromolecules 48(13): 4294–4300.

Ganesh, V., Sudhir, V.S., Kundu, T. and Chandrasekaran, S. 2011. 10 years of click chemistry: synthesis and applications of ferrocene-derived triazoles. Chemistry–An Asian Journal 6(10): 2670–2694.

Hu, M.L., Azad, M.A., Habibi, B., Rouhani, Olyaei, H.M.B., Liu, K.G. et al. 2020. Electrochemical applications of ferrocene-based coordination polymers. ChemPlusChem 85(11): 2397–2418.

Huo, M., Yuan, J., Tao, L. and Wei, Y. 2014. Redox-responsive polymers for drug delivery: from molecular design to applications. Polymer Chemistry 5(5): 1519–1528.

Hwang, T.L., Batabyal, D., Knutson, N. and Wikström, M. 2021. Use of the 2D 1H-13C HSQC NMR methyl region to evaluate the higher order structural integrity of biopharmaceuticals. Molecules 26(9): 2714.

Ifkovits, J.L. and Burdick, J.A. 2007. Photopolymerizable and degradable biomaterials for tissue engineering applications. Tissue Engineering 13(10): 2369–2385.

Jawaria, R., Khan, M.U., Hussain, M., Muhammad, S., Sagir, M., Hussain, A. et al. 2021. Synthesis and characterization of ferrocene-based thiosemicarbazones along with their computational studies for potential as inhibitors for SARS-CoV-2. Journal of the Iranian Chemical Society: 1–8.

Kim, C.R., Uemura, T. and Kitagawa, S. 2016. Inorganic nanoparticles in porous coordination polymers. Chemical Society Reviews 45(14): 3828–3845.

Kim, J., Baek, K.J., Yu, S., Yang, H.S., Khaliq, N.U., Choi, W. et al. 2023. Ferrocene-based acrylate copolymer multilayers with efficient antifouling and electrochemical redox properties. Electrochimica Acta 463: 142824.

Kovalska, V., Cherepanov, V., Losytskyy, M., Chernii, S., Senenko, A., Chernii, V. et al. 2014. Anti-fibrillogenic properties of phthalocyanines: Effect of the out-of-plane ligands. Bioorganic & Medicinal Chemistry 22(24): 6918–6923.

Krasavin, M., Lukin, A., Bagnyukova, D., Zhurilo, N., Zahanich, I., Zozulya, S. et al. 2016. Free fatty acid receptor 1 (GPR40) agonists containing spirocyclic periphery inspired by LY2881835. Bioorganic & Medicinal Chemistry 24(21): 5481–5494.

Larik, F.A., Saeed, A., Fattah, T.A., Muqadar, U. and Channar, P.A. 2017. Recent advances in the synthesis, biological activities and various applications of ferrocene derivatives. Applied Organometallic Chemistry 31(8): e3664.

Li, M., Li, J., Zhang, T., Zhao, Q., Cheng, J., Liu, B. et al. 2017. Syntheses, toxicities and anti-inflammation of H2S-donors based on non-steroidal anti-inflammatory drugs. European Journal of Medicinal Chemistry 138: 51–65.

Li, Y., Zhao, L., Liu, M., Zeng, X., Zhao, Y. and Zhou, B. 2024. Multifunctional polyoxovanadate-based covalent organic polymer for bacteria-infected wound therapy. ACS Applied Nano Materials 7(23): 26928–26940.

Lin, J., Yang, H., Zhang, Y., Zou, F., He, H., Xie, W. et al. 2023. Ferrocene-based polymeric nanoparticles carrying doxorubicin for oncotherapeutic combination of chemotherapy and ferroptosis. Small 19(2): 2205024.

Madruga, E.L. 2002. From classical to living/controlled statistical free-radical copolymerization. Progress in Polymer Science 27(9): 1879–1924.

Mahapatra, D.K., Bharti, S.K. and Asati, V. 2015. Anti-cancer chalcones: Structural and molecular target perspectives. European Journal of Medicinal Chemistry 98: 69–114.

McNicholas, B.J. 2020. New Electrolytic Media and Methods for Energy Storage and Conversion, California Institute of Technology.

Mcternan, C.T. 2018. Artificial Molecular Machines, The University of Manchester (United Kingdom).

Merchant, S.A., Tran, T.O., Meredith, M.T., Cline, T.C., Glatzhofer, D.T. and Schmidtke, D.W. 2009. High-sensitivity amperometric biosensors based on ferrocene-modified linear poly (ethylenimine). Langmuir 25(13): 7736–7742.

Niida, A., Yamaki, Y.K., Asakawa, T., Ishimura, Y., Fujita, H., Matsumiya, K. et al. 2018. Antiobesity and emetic effects of a short-length peptide YY analog and its PEGylated and alkylated derivatives. Bioorganic & Medicinal Chemistry 26(3): 566–572.

Nikitin, K., Bunz, H.M., Guiry, P.J. and McGlinchey, M.J. 2019. A mechanistic rationale for the outcome of Sonogashira cross-coupling of 9-bromoanthracene and ethynyltrimethylsilane: An unexpected product 4-(9-anthracenyl)-1, 3-bis (trimethylsilyl)-3-en-1-yne. Journal of Organometallic Chemistry 880: 1–6.

Niknam, E., Moaddeli, A. and Khazaei, A. 2020. Palladium anchored on guanidine-terminated magnetic dendrimer (G3-Gu-Pd): An efficient nano-sized catalyst for phosphorous-free Mizoroki-Heck and copper-free Sonogashira couplings in water. Journal of Organometallic Chemistry 923: 121369.

Odette, W.L. 2021. Amphiphilic Ferrocene Derivatives: Versatile Redox-Responsive Building Blocks of Functional Materials, McGill University (Canada).

Ogba, O.M., Warner, N.C., Leary, D.J.O. and Grubbs, R.H. 2018. Recent advances in ruthenium-based olefin metathesis. Chemical Society Reviews 47(12): 4510–4544.

Ornelas, C. 2011. Application of ferrocene and its derivatives in cancer research. New Journal of Chemistry 35(10): 1973–1985.

Ornelas, C. and Astruc, D. 2023. Ferrocene-based drugs, delivery nanomaterials and fenton mechanism: state of the art, recent developments and prospects. Pharmaceutics 15(8): 2044.

Patra, M. and Gasser, G. 2017. The medicinal chemistry of ferrocene and its derivatives. Nature Reviews Chemistry 1(9): 0066.

Pietschnig, R. 2016. Polymers with pendant ferrocenes. Chemical Society Reviews 45(19): 5216–5231.

Plamper, F.A. 2014. Changing polymer solvation by electrochemical means: basics and applications. Porous Carbons–Hyperbranched Polymers–Polymer Solvation: 125–212.

Quadri, C.C., Lalrempuia, R., Frøystein, N.A., Törnroos, K.W. and Roux, E.L. 2018. Steric factors on unsymmetrical O-hydroxyaryl N-Heterocyclic carbene ligands prevailing the stabilization of single stereoisomer of bis-ligated titanium complexes. Journal of Organometallic Chemistry 860: 106–116.

Rauf, U., Shabir, G., Bukhari, S., Albericio, F. and Saeed, A. 2023. Contemporary developments in ferrocene chemistry: physical, chemical, biological and industrial aspects. Molecules 28(15): 5765.

Rider, D.A. 2007. Synthesis, Self-Assembly and Applications of Amorphous, Department of Chemistry, University of Toronto.

Roy, G., Gupta, R., Sahoo, S.R., Saha, S., Asthana, D. and Mondal, P.C. 2022. Ferrocene as an iconic redox marker: From solution chemistry to molecular electronic devices. Coordination Chemistry Reviews 473: 214816.

Saleem, M., Yu, H., Wang, L., Abdin, Z.U., Khalid, H., Akram, M. et al. 2015. Review on synthesis of ferrocene-based redox polymers and derivatives and their application in glucose sensing. Analytica Chimica Acta 876: 9–25.

Savari, M. and Gharechaee, H. 2020. Application of the extended theory of planned behavior to predict Iranian farmers' intention for safe use of chemical fertilizers. Journal of Cleaner Production 263: 121512.

Scottwell, S.Ø. and Crowley, J.D. 2016. Ferrocene-containing non-interlocked molecular machines. Chemical Communications 52(12): 2451–2464.

Sha, Y., Shen, Z., Jia, H. and Luo, Z. 2020. Main-chain ferrocene-containing polymers prepared by acyclic diene metathesis polymerization: a review. Current Organic Chemistry 24(9): 1010–1017.

Sharma, A., Rana, R., Kumar, N., Gulati, H., Jyoti, K., Khanna, A. et al. 2025. Ferrocene-based compounds: promising anticancer and antimalarial agents in modern therapeutics. Medicinal Chemistry Research: 1–23.

Sharma, S. and Rathore, K. 2023. Stimuli responsive ferrocene-based supramolecular systems: New route for the design of smart devices. Journal of Organometallic Chemistry 997: 122787.

Singh, A., Lumb, I., Mehra, V. and Kumar, V. 2019. Ferrocene-appended pharmacophores: An exciting approach for modulating the biological potential of organic scaffolds. Dalton Transactions 48(9): 2840–2860.

Straube, A., Useini and Hawkins, E. 2025. Multi-ferrocene-based ligands: from design to applications. Chemical Reviews 125(6): 3007–3058.

Sun, R., Wang, L.I., Yu, H., Abdin, Z.U., ChenJin, Y. and Tong, H.R. 2014. Molecular recognition and sensing based on ferrocene derivatives and ferrocene-based polymers. Organometallics 33(18): 4560–4573.

Takahashi, S. and Anzai, J.-i. 2013. Recent progress in ferrocene-modified thin films and nanoparticles for biosensors. Materials 6(12): 5742–5762.

Van Staveren, D.R. and Metzler-Nolte, N. 2004. Bioorganometallic chemistry of ferrocene. Chemical Reviews 104(12): 5931–5986.

Wang, L. and Yu, H. 2018. Synthesis, Properties and Applications of Ferrocene-Based Derivatives, Polymers and Hydrogels, Springer.

Wang, Y., Astruc, D. and Ad-el-aziz, A.L. 2019. Metallopolymers for advanced sustainable applications. Chemical Society Reviews 48(2): 558–636.

Wei, J. and Diaconescu, P.L. 2019. Redox-switchable ring-opening polymerization with ferrocene derivatives. Accounts of Chemical Research 52(2): 415–424.

Woo, J., Na, Y., Choi, W., Kim, S., Kim, J., Hong, J. et al. 2021. Functional ferrocene polymer multilayer coatings for implantable medical devices: Biocompatible, antifouling, and ROS-sensitive controlled release of therapeutic drugs. Acta Biomaterialia 125: 242–252.

Wu, G.L., Tan, S., Tan, X., Guodong, C. and Yang, Q. 2025. Recent advances in ferrocene-based nanomedicines for enhanced chemodynamic therapy. Theranostics 15(2): 384.

Wu, J., Wang, L., Yu, H., Abdin, Z.U., Khan, R.U. and Haroon, M. 2017. Ferrocene-based redox-responsive polymer gels: Synthesis, structures and applications. Journal of Organometallic Chemistry 828: 38–51.

Xu, F., Li, H., Luo, Y.L. and Tang, W. 2017. Redox-responsive self-assembly micelles from poly (N-acryloylmorpholine-block-2-acryloyloxyethyl ferrocenecarboxylate) amphiphilic block copolymers as drug release carriers. ACS Applied Materials & Interfaces 9(6): 5181–5192.

Yan, Y., Zhang, J., Rend, L. and Tang, C. 2016. Metal-containing and related polymers for biomedical applications. Chemical Society Reviews 45(19): 5232–5263.

Yang, X., Wang, Y., Qi, W., Suabc, R. and He, Z. 2017. Bioorganometallic ferrocene-tripeptide nanoemulsions. Nanoscale 9(40): 15323–15331.

Yuan, Y., Subedi, L., Lim, D., Jung, J.K., Kim, S.Y. and Seo, S.Y. 2019. Synthesis and anti-neuroinflammatory activity of N-heterocyclic analogs based on natural biphenyl-neolignan honokiol. Bioorganic & Medicinal Chemistry Letters 29(2): 329–333.

Zhou, T., Zhang, S., Zhang, L., Jiang, T., Wang, H. and Huang, L. 2023. Redox ferrocenylseleno compounds modulate longitudinal and transverse relaxation times of FNPs-Gd MRI contrast agents for multimodal imaging and photo-Fenton therapy. Acta Biomaterialia 164: 496–510.

Chapter 8

Redox Sensitive Hyaluronic Acid Polymers for Drug Delivery Applications

Atefeh Derakhshani,[1] *Ali KarbalaeiMahdi,*[2] *Saeedeh Hasani*[3,*]
and *Najmeh Nematpour*[2]

1. Introduction

Biodegradable polymers and conjugates that respond to reductions have become an intriguing category of biomedical materials, which can be skillfully utilized for gene and drug delivery triggered intracellularly (Meng et al., 2009). These polymers typically have distinctive disulfide linkages either in the main chain, along the side chain, or within the cross-linker. While disulfide bonds are generally stable during circulation and in the extracellular environment, they can be quickly broken down within a time frame of minutes to hours when exposed to a reducing environment via thiol-disulfide exchange reactions (Gilbert, 1995; Raina and Missiakas, 1997).

Of late, various carrier systems are being developed and expanded, because they have various advantages. Advantages include:

- Capacity to increase drug solubility through encapsulation,
- Faster time to market and lower expense than improving the drugs themselves,

[1] Department of Tissue Engineering, School of Advanced Technologies in Medicine, Tehran University of Medical Sciences, Tehran, Iran.
[2] Department of Medical Nanotechnology, School of Advanced Technologies in Medicine, Tehran University of Medical Sciences, Tehran, Iran.
[3] Department of Nanotechnology & Advanced Materials, Materials and Energy Research Center (MERC), Karaj, Iran.
* Corresponding author: Hasani1369@gmail.com
DOI: 10.1201/9781003546672-8

- Capacity to decrease the required drug doses, as well as side effects,
- Ability to protect drugs from degradation.

Generally, an ideal multifunctional nanoplatform for treatment needs good biocompatibility, effective drug delivery and imaging, and targeting ability. Various kinds of nanoparticles were designed to optimize treatment strategies. Particularly, the studies and applications of self-assembled hyaluronic acid (HA) nanoparticles as anticancer therapeutic agents have attracted much attention recently (Huang et al., 2017; Huang et al., 2014; Yin et al., 2015; Yin et al., 2015; Zhang et al., 2014). However, many HA nanoparticles are restricted to applications in drug delivery for chemotherapy, which may show some deficiencies like endotoxicity or drug resistance and lead to side effects (Zhang et al., 2014).

Recent literature indicates that the molecular weight of HA and modified HA systems impact numerous cellular and molecular signaling pathways, as well as matrix remodeling and the maintenance of tissue homeostasis (Amorim et al., 2021). As a result, nanocarriers that utilize HA as the main targeting component have been extensively researched for drug delivery purposes (Lee et al., 2020). Coating nanoparticles, like liposome nanoparticles (Song et al., 2019), chitosan (CS) nanoparticles (Lallana et al., 2017), and metal-organic frameworks (Shu et al., 2018), with HA might also decrease cytotoxicity. Targeted cellular uptake of HA-containing systems in primary or metastatic tumors indicates the high antitumor efficacy and effectiveness of delivery systems (Yang et al., 2017). Simultaneous drug delivery and specific inhibition of cancer stem cells are other advantages of attaching HA to nanoparticles.

HA has a distinctive chemical composition that permits easy chemical alterations. Thiol-HA (HA-SH) is generated by modifying the carboxylate groups, with the thiol having the potential to be oxidized, forming disulfide linkages (Cao et al., 2019). The initial attachment of ligands with free thiol groups to the polymeric backbone of HA was first reported by Bernkop-Schnürch, paving the way for numerous pharmaceutical applications through functional customization (Griesser et al., 2018). By thiolating HA, properties like mucoadhesiveness, swelling ability, stability, and biocompatibility can be enhanced (Prestwich, 2011). The stabilization of thiol groups through crosslinking or preactivation that creates disulfide bonds leads to the development of more stable and efficient drug delivery systems, enhancing the characteristics of thiomers even further (Iqbal et al., 2012). In recent years, there has been considerable interest in redox-sensitive HA nanocarriers that incorporate disulfide bonds, primarily due to the marked differences in the reducing conditions present in intracellular and extracellular environments (Hou et al., 2023; Li et al., 2016; Jian Liu et al., 2018; Mahdieh et al., 2022; Wang et al., 2021). In this chapter, the initial focus will be on the modifications made to HA and their associated properties. The distinctive characteristics of functionalized HA enable their application as delivery vehicles targeting tumors and the cancer microenvironment. A summary of the delivery systems will be provided in the last part of the chapter.

2. Redox-Manipulating Delivery Systems

The tumor microenvironment can be leveraged to develop drug delivery systems that respond to particular stimuli, enabling the targeted release of therapeutics at designated locations. This approach enhances the efficiency of carrier delivery, increases the bioavailability of the drugs, and minimizes both toxicity and adverse effects (Hossen et al., 2019). In biological systems, glutathione (GSH, a tripeptide that includes cysteine) exists at varying levels in blood circulation (around 2 ~ 20 μM) and in the cytoplasm of mammalian cells (about 2 ~ 10 mM). However, the GSH concentrations in cancer cells were markedly elevated, reaching roughly 20 mM (Li et al., 2015). Glutathione exists in two distinct forms. The first is the reduced form, known as GSH, which is present in millimolar concentrations within the intracellular cytosolic compartment. The second form, the oxidized variant (GSSG), constitutes only a minor fraction of the total glutathione found within cells. GSH, recognized as a principal antioxidant within cells, plays a crucial role in the biosynthesis and metabolism that influence various intracellular mechanisms, including enzyme catalysis, drug detoxification, and cellular signal transduction. GSH plays a significant role in modulating the progression of cancer through various mechanisms. The phenomenon of oxidative stress is frequently associated with cancer advancement, which can be attributed to the elevated metabolic activity and the activation of signaling pathways linked to reactive oxygen species (Valko et al., 2006).

GSH, along with other intracellular antioxidants, plays a crucial role in safeguarding cells by mitigating oxidative stress. The redox state of intracellular thiols (regulated by GSH) is vital for modulating the mitochondrial permeability transition pore complex. Thiol oxidation may play a role in mitochondrial processes that result in tissue injury. Notably, disulfide is typically integrated into the GSH-responsive transport systems. The release of cargo is initiated by the thiol-disulfide exchange reaction that occurs between GSH and the vehicle containing disulfide bonds. Thiol–disulfide exchange reactions are quick and easily reversible, and they are crucial for sustaining the proper biological functions of living cells, such as the stabilization of protein structures, enzymatic activity, and redox cycles (Arnér and Holmgren, 2000; DM, 2003; Go and Jones, 2008). Thiol oxidation is generally a slow process under physiological conditions (pH 7.4) since it requires the thiol groups to be deprotonated, forming thiolate ions ($R\text{-}S^{-}$) that subsequently react with molecular oxygen to produce reactive radical species. Because thiols have a pKa of approximately ~ 8–10, basic conditions (pH > 8.5) are necessary to promote the forward reaction (Peng et al., 2013). Numerous chemical bonds exhibit sensitivity to redox conditions, including the disulfide bond (-S-S-), monothioether bond (-S-), platinum-oxygen conjugate bond (-Pt-O-), diselenide-conjugated bond (-Se-Se-), selenium-nitrogen conjugate bond (-Se-N-), and monomolecular selenium bond (-Se-). Disulfide bonds have been extensively utilized in the creation of reduction-responsive drug delivery systems aimed at cancer treatment (Li et al., 2021).

Redox homeostasis refers to the mechanism through which the intracellular microenvironment maintains a balanced and stable equilibrium between oxidants and antioxidants (Chen et al., 2021; Kou et al., 2020). The variation in GSH levels can be leveraged to enhance the release of drugs from micelles through a redox-sensitive conjugate. The difference in redox potential may serve as a viable trigger for the release of pharmaceuticals. Elevated concentrations of the antioxidant GSH within tumor cells are crucial for the neutralization of reactive oxygen species (ROS) and the preservation of redox balance (Fu et al., 2021). This characteristic may contribute to the enhanced resistance of tumorigenic cells to therapies that rely on ROS (Meng et al., 2024). The development of the GSH-sensitive delivery platforms was crucial for the release of the cargo, subsequently leading to the depletion of GSH within the tumor cells.

Redox-sensitive carriers are capable of detecting elevated intracellular concentrations of GSH and subsequently releasing their payload in a controlled manner. Stimuli-responsive nanocarriers represent a highly promising approach in the field of drug delivery across various dimensions. A significant benefit of triggered nanomedicine lies in its ability to prevent the early release of therapeutic agents; specifically, the drug remains encapsulated within the nanocarrier throughout systemic circulation until it arrives at the intended target site (Meng et al., 2024). The release of the cargo can be induced in a controlled manner through the application of either internal or external stimuli, thereby elevating the concentration of the drug at the affected site and improving therapeutic effectiveness. In addition, a diverse array of stimuli is available for inducing drug release, tailored to the specific type, anatomical site, and pathological characteristics of various cancers. In some instances, a singular nanoplatform may exhibit responsiveness to multiple stimuli, thereby facilitating the development of a multifunctional anticancer nanomedicine (Meng et al., 2024).

External stimuli utilized for the regulation of nanoscale drug delivery typically encompass ultrasound, temperature variations, light exposure, and magnetic fields. These stimuli typically do not depend on the characteristics of the tumor pathology and its surrounding microenvironment. Internal stimuli, often referred to as endogenous or biological stimuli, primarily influence the damaged tissues, thereby enhancing the specificity of biomedical applications. Internal stimuli for the regulation of drug delivery systems primarily encompass particular pH levels, elevated concentrations of GSH resulting from oxidative stress in tumors and other specialized tissues, as well as the presence of over-expressed enzymes such as proteases, glycosidases, and phospholipases. These stimuli typically depend on the characteristics of the tumor pathology and the surrounding microenvironment (Fig. 1). Consequently, it is essential to develop optimal stimuli-responsive carriers that can react to specific internal signals, subsequently undergoing gradual disintegration to release their encapsulated cargo (Raza et al., 2019).

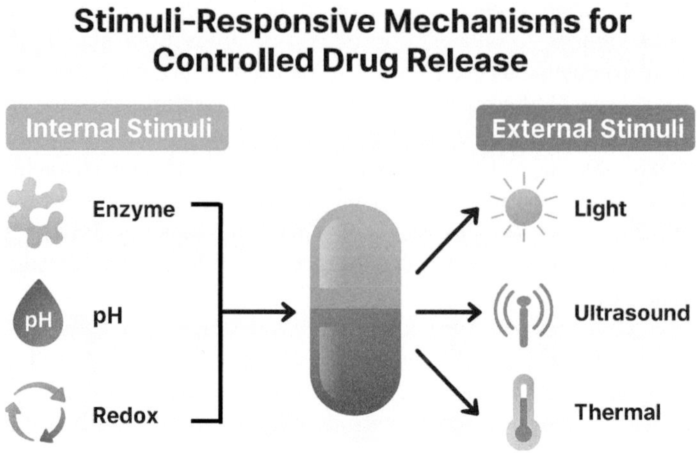

Fig. 1. Diagram of stimuli controlled release.

3. Chemical Modification Hyaluronic Acid (HA)

HA, which is a combination of hyaloid (vitreous) and uronic acid (UA), was first extracted in 1934 from the vitreous humor of bovine eyes by researchers Meyer and Palmer (Meyer and Palmer, 1934). It is a naturally occurring linear, non-protein, non-sulfated glycosaminoglycan (Fig. 2.a) that consists of repeating disaccharide units of D-glucuronic acid and N-acetyl-D-glucosamine, and it serves as a primary component of the extracellular matrix (ECM) in various biological tissues. The length of the linear polymer chain can vary greatly and can attain a molecular size of up to 10^7 Da (Allawadhi et al., 2022). In the past few decades, there has been considerable progress in both native HA and modified HA systems due to their beneficial biochemical, biophysical, viscoelastic, and tissue remodeling properties (Kotla et al., 2021; Toole, 2004). Combining HA with other substances can produce a synergistic effect, resulting in enhanced properties for drug loading and controlled release. As an example, poly(propylene glycol) (PPG) exhibits temperature-sensitive properties, and a hybrid thermogel composed of HA grafted with PPG can extend the release duration of an incorporated medication (Kim et al., 2022). Carriers composed of HA and chitosan exhibit superior drug encapsulation and demonstrate improved stability when exposed to salt ion concentrations and proteins found in plasma (Li et al., 2021; Turcsányi et al., 2020).

Chemical modification of HA can be achieved through two primary approaches: crosslinking and conjugation techniques. The chemical modifications of HA (Fig. 2.b) have predominantly been executed at various locations, such as the carboxyl groups of β-d-glucuronic acid (GlcUA), the hydroxyl groups of the N-acetyl-β-d-glucosamine (GlcNAc) residue, as well as thiolation of carboxylic groups (Fig. 2c) and the reducing terminus of the disaccharide repeating unit. The generation of primary amino groups on the hyaluronic acid (HA) polymer chain can also be achieved through the deacetylation of N-acetylglucosamine (GlcNAc), facilitating the chemical conjugation of acid derivatives through the formation of

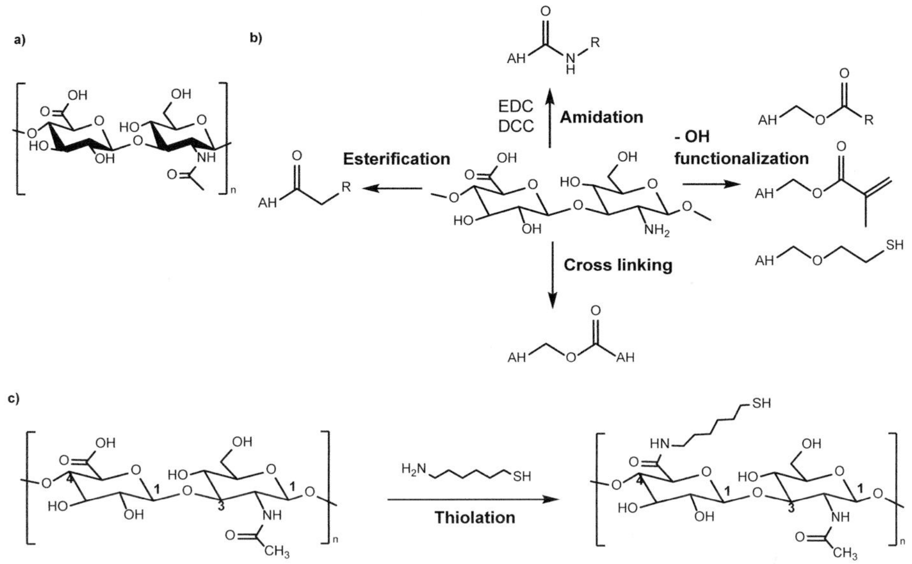

Fig. 2. Schematic illustrations of: (a) HA disaccharide unit chemical structure, (b) Chemical alterations of HA, such as esterification, amidation, and crosslinking involving the carboxylic groups of individual monomers, as well as various hydroxyl group functionalizations, (c) thiolation of carboxylic groups of HA.

amide bonds (Hou et al., 2023). The carboxylic acid group of HA is often subjected to reactions with primary amine derivatives through carbodiimide-mediated processes in aqueous environments (Mahdieh et al., 2022). With a biological half-life of up to 24 hours, HA is consistently broken down in the body by the hyaluronidase enzyme and through its internalization by CD44 receptors on the cell surface (Aruffo et al., 1990; Brown et al., 1991). In addition to enzymatic breakdown, HA may be degraded through various methods, including acidic or alkaline hydrolysis, thermal degradation, and breakdown by oxidizing agents (Stern et al., 2007). Chemical modifications and crosslinking are essential to extend the degradation period and enhance the mechanical stability both *in vivo* and in response to environmental factors (Kafedjiiski et al., 2007).

HA is found in ionized or salt forms when at physiological pH, for example, sodium hyaluronate. Due to the presence of carboxylic groups on each monomer, the molecule demonstrates anionic properties, leading to a significant degree of hydrophilicity. Furthermore, the abundant negatively charged hydroxyl groups in HA can draw in and hold a significant amount of water molecules (Gallo et al., 2019). The pKa values for carboxyl groups in HA range from 3 to 4; at a pH of 7, these groups are ionized, making the hyaluronan molecule a polyanion that interacts with cations. HA is an extremely water-attracting polymer capable of absorbing significant quantities of water and increasing its volume by up to 1000 times, creating a loose hydrated structure (Laurent and Fraser, 1992). The biophysical and mechanical properties of HA can be adjusted and are influenced by various modified forms of HA, including bioconjugate systems. Biopolymers, such as HA with biochemical response mechanisms, offer crucial insights for creating improved modified-HA-based conjugates or scaffolds. This development necessitates exploring different synthesis techniques to establish uniform modified-HA-based systems that offer improved pharmacological and therapeutic advantages (Kotla et al., 2023).

When combined with nanographene oxide (Wen et al., 2012), HA can create a protective layer on the nanosheet surface. This layer may impede the release of drugs loaded into the nanocarrier, significantly reducing the therapeutic effectiveness of the nanosheets. Consequently, it becomes crucial to employ specific techniques, such as a redox-responsive mechanism, to eliminate the HA barrier and enhance drug release. The redox environment within tumor cells is a critical factor that can influence how tumors respond to certain chemotherapy drugs and radiation treatments. HA-drug conjugates have been effectively created utilizing adipicdihydrazide (ADH) (Luo and Prestwich, 1999; Luo et al., 2000) or amino acids as linkers (Luo et al., 2000). Nevertheless, nearly all of these conjugates exhibit a comparatively slow release of the drug, which could considerably hinder the effectiveness of the encapsulated drug. At present, certain characteristics specific to the tumor microenvironment have been utilized to address this problem. Researchers have discovered that a comprehensive understanding of tumors and their microenvironments reveals a strong correlation between tumor proliferation, invasion, and metastasis, and the atypical features of the tumor microenvironment (Zhang et al., 2019). The encapsulation of pharmaceuticals within redox-sensitive polymer nanocarriers has emerged as the most extensively researched approach in contemporary studies (Akita et al., 2013). Generally, the disulfide bond is a frequently utilized connection in redox-sensitive systems (Dan et al., 2015), which can be broken down into the respective thiols in a reducing atmosphere or via a disulfide-thiol exchange (Tsarevsky and Matyjaszewski, 2002). Diselenide bonds, akin to disulfide bonds, may be disrupted in a redox environment pertinent to tumors (Chen et al., 2022; Ma et al., 2010). Block copolymers that incorporate diselenide bonds exhibit the ability to respond to γ-rays, suggesting that these diselenide bonds may facilitate the integration of chemotherapy and radiotherapy (Kieliszek and Lipinski, 2020; Kim et al., 2019). Furthermore, it has been documented that selenium compounds may assist in the treatment of COVID-19 (Kieliszek and Lipinski, 2020). The utilization of diselenide bonds was studied (Kim et al., 2019); tetra acid (TA) was activated with EDAC/NHS system and then reacted with selenocystamine to produce TA-selenocystamine conjugates. To initiate one of the carboxylic groups of chlorin e6 (Ce6), a solution containing Ce6 was combined with EDAC and NHS in equal molar proportions. The activated carboxylic acid of Ce6 was then conjugated with the amine end group of TA-selenocystamine conjugates. Subsequently, the HA-selenocystamine conjugates underwent a reaction with TAseseCe6 tetramer conjugates to generate HAseseCe6TA conjugates, as illustrated in Fig. 3. The characteristic signals of HA, selenocystamine, Ce6, and TA were identified in the range of 1.6 to 7.0 ppm. Specifically, the acetyl proton of HA was observed at 1.8 ppm, the ethylene proton of TA appeared between 2.2 and 2.5 ppm as well as 3.8 and 4.0 ppm, while the protons of Ce6 were detected at 1.6 ppm and between 6.0 and 7.0 ppm (Fig. 3).

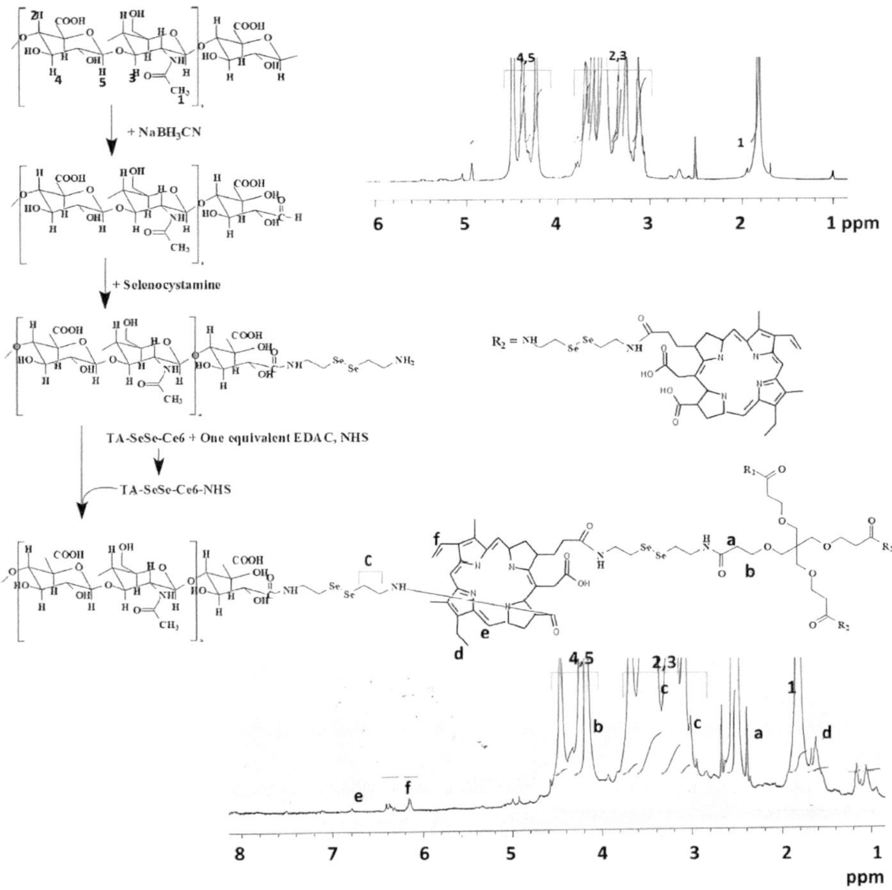

Fig. 3. Synthesis method and ¹H NMR spectra of HAseseCe6TA (Kim et al., 2019).

4. Properties of Modified Hyaluronic Acid

4.1 Biocompatibility

Biocompatibility is a crucial aspect of formulation development, relevant to both academic and industrial contexts since new formulations must not lead to harmful effects following administration. In various studies conducted by Laffleur and colleagues, thiolated HA showed no impact on the viability of Caco-2 cells (de Sousa et al., 2016; Laffleur et al., 2015a; Laffleur et al., 2014; Laffleur et al., 2015b). Thiolated HA did not show any cytotoxicity towards primary human fibroblasts either (Serban et al., 2008). Furthermore, a different study demonstrated improved absorption of fibronectin through a higher level of sulfonation (Köwitsch et al., 2016). This mechanism facilitates an increased attachment of cells to the surface of glycosaminoglycans. Furthermore, the extent of thiolation was examined, demonstrating that an increase in thiolation correlates with a decrease in the biocompatibility of glycosaminoglycans. Metabolic activity and cellular proliferation were observed to be enhanced to an elevated threshold of thiolation, by Clausen et al. (Clausen et al., 2002b). The composition of hydrogels is essential in identifying suitable compounds for particular applications, as lower molecular weight HA has been shown to promote optimal cell survival.

4.2 Properties that Enhance Penetration

The permeation enhancement is mainly the result of reduced GSH, as mentioned earlier. The thiol groups present in the polymer facilitate the conversion of oxidized glutathione into its reduced form through a thiol-exchange reaction (Clausen et al., 2002). The enhancement of permeation observed with HA-SH is primarily a result of the previously mentioned effect in most cases. The permeation rate of curcumin, for example, was observed to be increased by a factor of 4.4 when thiolated HA was employed in comparison to unmodified HA (Laffleur et al., 2017). In contrast, a study examining insulin permeation through a cell monolayer revealed that the permeation rate of insulin from the HA-SH gel was reduced in comparison to that of the unmodified gel. The presence of a greater number of thiol groups within the polymer backbone resulted in a reduced rate of permeation. In this context, the establishment of disulfide bonds between insulin and HA-SH was more significant than the permeation-enhancing properties of the thiolated polymer (Li et al., 2012).

4.3 Controlled Release

The modulation of the release of active pharmaceutical ingredients embedded within thiolated hyaluronic acid systems can be regulated by various factors, including alterations in structure and pH levels. Censi and colleagues (Censi et al., 2010) created *in situ* forming hydrogels through a combination of tandem thermal gelling and the Michael addition reaction, which serve as a carrier system for the controlled release of the peptide bradykinin. The thiolated HA system demonstrates significant potential for the delivery of both peptide and protein therapeutics, as the release kinetics of gels containing bovine serum albumin can be modified through adjustments to the polymer's chemical structure (Griesser et al., 2018). Fu et al. introduced an innovative release system that is responsive to both pH and reduction, employing doxorubicin hydrochloride as the active agent (Fu et al., 2015). An enhanced release of doxorubicin was observed at an acidic pH of 5, particularly in the presence of elevated concentrations of GSH.

4.4 Mucoadhesive Properties

The enhanced mucoadhesive characteristics of HA-SH primarily stem from the formation of disulfide bonds between the sulfhydryl groups present in the polymer backbone and the cysteine-rich residues found in mucus. Unmodified mucoadhesive polymers are limited to establishing ionic interactions, hydrogen bonds, and electrostatic bonds with the mucosal surface (Oh et al., 2015). The rotating cylinder method is among the most extensively researched and practical approaches for comparing and assessing the *in vitro* mucoadhesiveness of both unmodified and modified polymers (Bernkop-Schnürch et al., 2003). Kafedjiiski et al. demonstrated that the adhesion time was extended by a factor of 6.5 when comparing HA-cysteine ethyl ester (HA-Cys) to HA, utilizing the rotating cylinder method for their analysis (Kafedjiiski et al., 2007). The results are derived from the establishment of covalent bonds between the thiol groups of thiolated HA and the cysteine-rich regions present in mucin glycoproteins (Shinkar et al., 2012).

4.5 Cohesive Properties

The ability to swell significantly influences the adhesive and cohesive characteristics of thiolated HA. The swelling capacity is typically assessed by evaluating water absorption over a period. Several research teams have validated that HA-SH exhibits enhanced characteristics compared to HA regarding its capacity for water absorption. HA-cysteine ethyl ester demonstrated a swelling capacity that was 3.5 times greater than that of the unmodified polymer (Laffleur et al., 2015).

4.6 Stability against Degradation

The swelling capacity, as previously noted, plays a crucial role in the mucoadhesive properties of thiolated HA. The stability of the substance is adversely affected by significant swelling, as this increases the surface's susceptibility to hyaluronidases. In light of the delicate nature of hydrogel applications, *in vivo* investigations into the stability and degradation of HA-SH hydrogels are of significant interest. Chemical modifications, such as the thiolation of HA, are expected to enhance the polymer's resistance to enzymatic degradation (Liu et al., 2005). Nevertheless, the biocompatibility of the modified HA must remain intact. Laffleur et al. (Laffleur et al., 2015) demonstrated that chemical modification significantly influences stability, as the thiolation process enhances the polymer's resilience through the formation of both intra- and inter-molecular disulfide bonds, thereby increasing its resistance to degradation.

5. Redox-Sensitive Hyaluronic Acid Functionalized Carriers

Cancer is a complex disease with various etiological factors identified across different research studies. It arises from abnormalities in cellular regulatory mechanisms or alterations in genetic material, which can lead to mutations or the overexpression of specific genes. These changes contribute to unregulated cellular growth and proliferation, resulting in the development of tumors (Chu et al., 2024; Tuffaha et al., 2024).

Chemotherapy is a key strategy in cancer treatment, but it faces major challenges such as non-selective toxicity, poor water solubility, and rapid clearance by the reticuloendothelial system (RES) (Eslami et al., 2024). These issues complicate drug formulation and dosing. However, nano-scale polymeric drug delivery systems (DDSs) offer a promising solution. With their nanoscale dimensions, unique properties, and core/shell structures, these systems effectively solubilize hydrophobic drugs, protect them from degradation, extend their circulation time, and enhance accumulation in tumors. This innovative approach significantly improves the bioactivity and bioavailability of cancer therapies, leading to more effective treatments (Hari et al., 2023).

In drug delivery research, HA has emerged as a promising carrier as it has several useful properties, including (i) its ability to biodegrade; (ii) its compatibility with biological systems; (iii) the simplicity of its chemical modification; (iv) its capacity for high drug loading; and (v) its natural targeting capabilities, which stem from specific interactions with receptors like CD44 or hyaluronan receptors involved in endocytosis (HARE) (Raza et al., 2019). Although promising outcomes have been observed in various *in vitro* and *in vivo* studies, there remains a need for a deeper understanding of the pharmacokinetics and pharmacodynamics of chemically modified or HA-conjugated formulations before their application in therapeutic contexts. Although initiatives are underway in this area, additional research is required to determine the extent of substitution of drug moieties on the hyaluronic acid (HA) chain/system and its implications for stability, pharmacokinetics, release, and other related factors.

The development of nano-scale drug delivery systems for anticancer therapies has emerged as a significant area of investigation in the realm of targeted chemotherapy for tumors. Targeted drug delivery systems utilizing nanoparticles, which integrate both passive and active targeting mechanisms for cancer therapy, have been the focus of extensive research and development over the last thirty years (Yu et al., 2012). This section will explore the significance of HA-modified delivery systems and their roles in therapeutic interventions for various diseases, specifically cancer treatment. The outlook is summarized in Fig. 4.

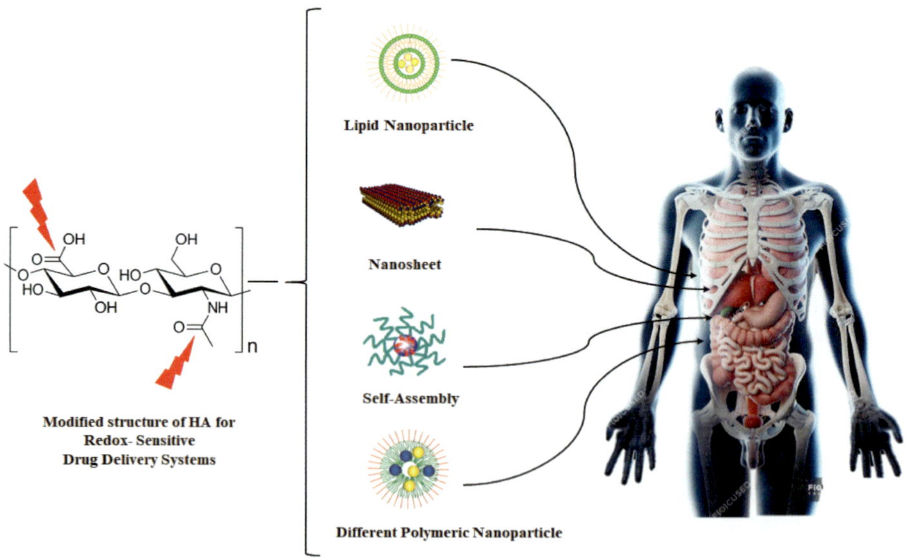

Fig. 4. Illustration of redox-sensitive hyaluronic acid delivery systems.

5.1 Nanoparticles

Polymeric micelles that self-assemble represent a type of nanocarrier system with possible uses in cancer treatment. This approach has the potential to significantly boost the effectiveness of anticancer medications by increasing their solubility in water, extending their circulation time in the bloodstream, and enhancing their selective targeting of tumor tissue through the enhanced permeability and retention (EPR) effect. Among the various polymeric micelles, polymer-drug conjugates (PDCs) stand out for their exceptional properties (Gupta et al., 2023). Their inherent amphiphilic characteristics enable them to self-assemble into nano-micelles in aqueous environments, creating an effective method for encapsulating hydrophobic anticancer drugs. PDCs feature micro-hydrophobic pockets that provide excellent compatibility with the drugs they contain, allowing for significantly higher drug-loading capacities and improved resistance to dilution in body fluids (Yin et al., 2015). This reduces the risk of drug leakage during circulation, surpassing the performance of conventional polymeric micelles. However, many PDCs based on leading antitumor agents like taxanes, camptothecin, and doxorubicin have not achieved optimal efficacy due to challenges in intracellular trafficking and drug release (Lv et al., 2014). The stable ester bond linkages in these conjugates degrade slowly *in vivo*, and strong hydrophobic interactions hinder effective release within tumor cells, ultimately limiting their anticancer effectiveness. To overcome these obstacles, there is a strong need for a PDC micellar system that facilitates active intracellular targeting and stimuli-triggered drug release. By effectively encapsulating drug molecules while enabling selective release in the intracellular environment, this innovative approach can significantly improve the formulation and delivery of anticancer therapies, addressing toxicity concerns without compromising drug release efficiency. The hydrophobic interior of the micelles serves as a storage compartment for water-insoluble pharmaceuticals, while the hydrophilic exterior functions as a protective layer that enhances the stability of the micelles in aqueous environments (Savic et al., 2003). An effective strategy for facilitating active tumor targeting involves the incorporation of targeting ligands, such as HA, onto the surface of micelles through a range of conjugation techniques. Mahdieh and her colleagues present a novel micellar waterborne polyurethane system that is targetable, clickable, and capable of cleaving tumors, designed specifically for intracellular drug delivery (Mahdieh et al., 2022).

HA can form nanoparticles through self-assembly when it is chemically linked to hydrophobic groups such as 5β-cholanic acid, 1-pyrenebutanamide, and octadecylamine (Zhang et al., 2014). However, traditional polymeric micellar delivery systems often encounter low cellular uptake within cancer cells due to the steric shielding caused by their very hydrophilic outer shells. Additionally, the slow degradation of micellar structures results in a prolonged release of the drug over an extended duration (Li et al., 2012). Recently, researchers have developed redox-sensitive micelles that can initiate a rapid release of encapsulated medications in reaction to the intracellular redox potential (Meng et al., 2009).

HA-based systems employed as drug delivery agents, along with nano-systems that maintain physiologically stable conditions, significantly improve the antitumor and targeted therapeutic effects in malignant tumors (Luo et al., 2000). Novel redox/pH dual-responsive nanoparticles exhibiting reversible surface charge were synthesized utilizing HA-SH and CS (Fig. 5.a) to facilitate the controlled release of the anticancer agent doxorubicin (Xia et al., 2021). Nanocarriers that respond to stimuli were synthesized through a three-phase procedure, Fig. 5.b. In the initial phase, cysteamine dihydrochloride reacted with the carboxylic groups of HA through its primary amine groups, utilizing the EDC/HOBt coupling method. DTT was employed to cleave the disulfide bonds present in the cysteamine-modified hyaluronic acid during the subsequent phase. The third phase involved the electrostatic self-assembly and subsequent oxidation of the thiol groups. The surface charges of the nanoparticles exhibited reversibility when subjected to varying mass ratios of HA-SH and CS. The stability of nanoparticles in physiological environments was enhanced by the incorporation of additional chemical cross-linking involving the HA-SH thiol groups. HA-SH/CS nanoparticles exhibit numerous advantageous characteristics, including remarkable biocompatibility and biodegradability, targeted interaction with CD44 via HA, sufficient drug-loading capacity, and prompt drug release triggered by intracellular pH and reducing conditions, Fig. 5.c.

For enhancing antitumor efficacy, an HA-modified hybrid nanocomplex HA-diselenide bond-carboxylic aid-poly (amidoamine) was constructed to deliver siRNA into tumor cells via a redox-mediated intracellular disassembly (Chen et al., 2022). Cationic polymer-based nonviral gene delivery has gained recognition as a promising alternative to viral vectors due to several advantages, including straightforward manufacturing processes, the absence of restrictions on DNA size, reduced inflammatory responses, and minimal immunogenicity (Sun et al., 2008). Due to the presence of terminal amino groups, poly (amidoamine) (PAMAM) molecules exhibit a positive charge in a physiological pH environment, enabling them to interact with the negatively charged phosphate groups found in RNAs. The synthesized nanocomplex demonstrated high transfection efficiency, attributed to its effective internalization through a clathrin-dependent pathway and its remarkable ability to escape from endosomes. Cheng et al. similarly observed that polyethyleneimine (PEI) nanoparticles linked with diselenide groups and loaded with DNA exhibit enhanced transfection efficiency when compared to conventional PEI, while also demonstrating a degradation profile that is sensitive to redox conditions (Cheng et al., 2012).

Photodynamic therapy (PDT) has the potential to cause cellular damage or apoptosis via ROS produced by photosensitizers (PS). The extent of ROS generation plays a critical role in determining the efficacy of PDT. The synthesized nanoparticles demonstrated an enhanced anti-tumor efficacy through the inhibition of microtubule depolymerization, disruption of the cell cycle, and the production of ROS. In research (Wang et al., 2021) to this end, molecules were altered through the incorporation of hydrophobic docosahexaenoic acid (DHA) and ce6, utilizing cysteamine (cys) as a linking agent. This modification aimed to produce amphiphilic HA derivatives capable of physically encapsulating docetaxel. Finally, the results indicate the increased antitumor and targeted effect of this system in cancer cells. In another study about redox-responsive theranostic, Yang et al. developed a nanoparticle

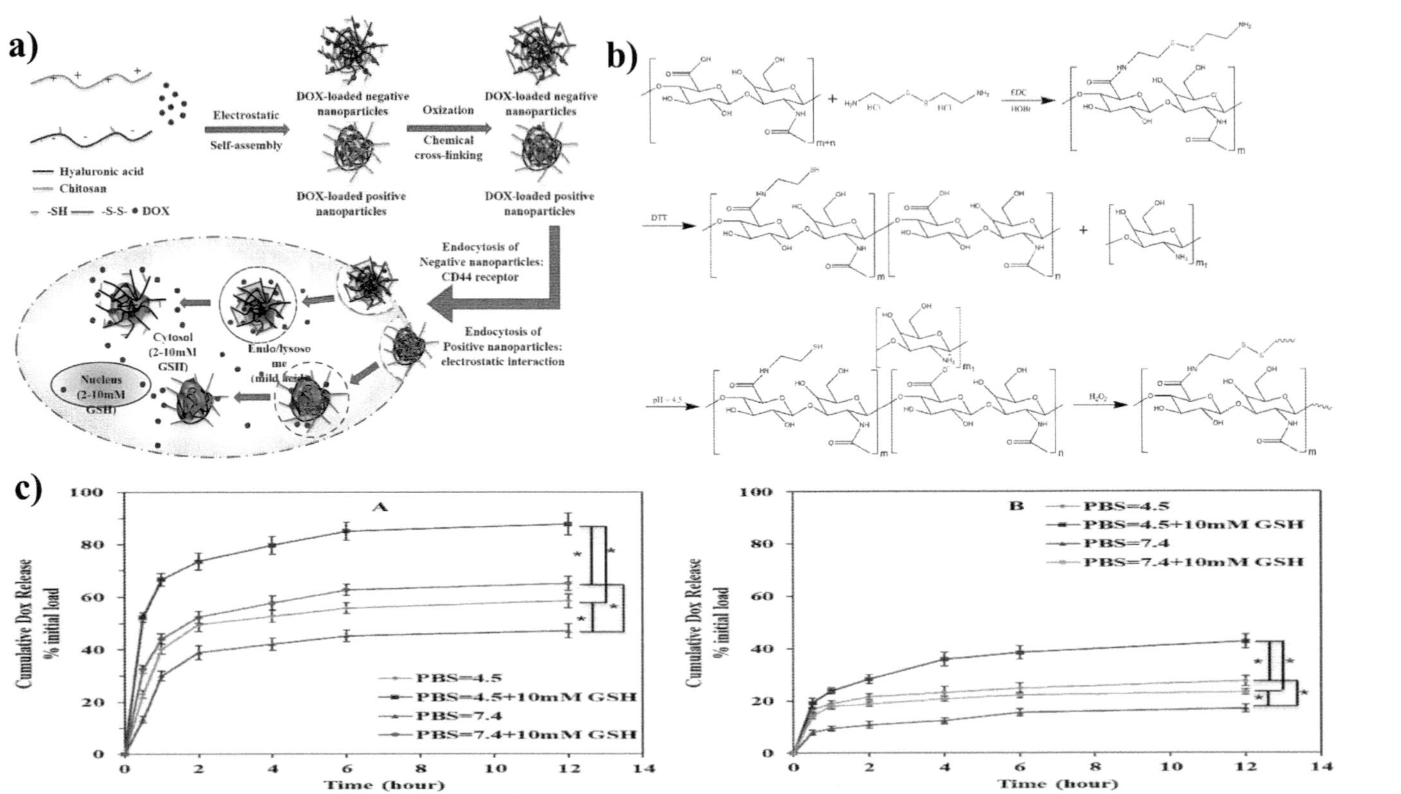

Fig. 5. (a) Schematic representation of the synthesis, redox responsiveness, and pH-sensitive drug release mechanisms of HA-SH-CS nanoparticles. (b) Three synthesis steps for the preparation of HA-SH/CS nanoparticles. (c) Redox/pH dual-responsive DOX release profiles from DOX-loaded HA-SH/CS nanoparticles under different conditions of pH and GSH concentrations. (A) Negative nanoparticle. (B) Positive nanoparticles. There was a significant difference between different conditions for both nanoparticles (Xia et al., 2021).

composed of a HA and poly(“-caprolactone”) copolymer that is responsive to GSH, which encapsulates doxorubicin and superparamagnetic iron oxide (SPIO). The application of elevated concentrations of GSH resulted in the cleavage of disulfide bonds within these nanoparticles, leading to the release of encapsulated doxorubicin and SPIO. The released SPIO can be employed in magnetic resonance imaging, whereas doxorubicin serves a therapeutic role in chemotherapy, thereby facilitating the integration of imaging techniques with chemotherapeutic interventions (Yang et al., 2019).

In a recent study (Hou et al., 2023), researchers successfully developed HA-modified porous silica ($pSiO_2$) nanocarriers aimed at enhancing the efficiency of drug delivery and facilitating combination therapy. $pSiO_2$ nanocarriers exhibit stable porous structures and high loading capacity, while $pSiO_2$/HA nanocarriers offer the combined advantages of HA-based carriers and $pSiO_2$ nanoparticles. The synthesis of $pSiO_2$ nanocarriers involved a two-step process, after which Ag2S QDs were incorporated into the porous structure of the $pSiO_2$ vessels, endowing them with photothermal capabilities. The $pSiO_2$ nanocarriers exhibit dimensions of 30 nm, featuring extensive channels and substantial loading capacity (29.3%). To attach HA, a three-step reaction process was employed to conjugate a responsive linker containing an alkyl amine and disulfide bond onto the surface of Ag2S/$pSiO_2$ nanocarriers.

5.2 Gels

The redox-sensitive HA hydrogel is a promising material for drug delivery, diagnostic wound monitoring, and tissue regeneration. Its ability to respond to biochemical signals like GSH, track wound healing, and release drugs selectively makes it valuable for personalized medicine (Gao et al., 2019). In an investigation aimed at treating and delivering therapeutic agents in a chronic wound environment, Gao and co-workers described the facile, one-step synthesis of an HA-based chemical hydrogel that is spatially responsive to the presence of GSH. Aminoethyl disulfide (AED) is used as a dynamic covalent crosslinker that is sensitive to the presence of GSH. The conjugation of AED with GSH results in crosslinker cleavage, increased hydrogel swelling, and a visible change in hydrogel dimensions. The hydrogel was determined to be non-cytotoxic and supported fibroblast growth and proliferation over a week (Gao et al., 2019).

Nanogels (NGs) with excellent biocompatibility, high water content, and exceptional protein compatibility have attracted much interest in protein delivery (Li et al., 2015; Tahara and Akiyoshi, 2015; Zhang et al., 2015). Utilizing the unique characteristics of the naturally occurring tumor microenvironment or intracellular milieu, various stimuli-responsive nanogels have been investigated to achieve targeted protein release at specific sites (Cheng et al., 2015; Lu et al., 2014). Kurisawa et al. demonstrated that a ternary nanogel composed of hyaluronic acid, green tea catechin, linear polyethylenimine, and granzyme B (GrB) successfully facilitated the targeted intracellular delivery of GrB to HCT-116 cancer cells *in vitro* (Liang et al., 2016). To enhance the typically low stability of physical nanogels, various chemical cross-linking techniques have been utilized, including radical polymerization, the Michael addition reaction, copper-mediated click reactions, and amidation, to produce nanogels with improved stability (Jiang et al., 2014; Oh et al., 2008). In a study about HA-modified and cross-linked nanogels, Li et al. presented findings on redox-sensitive and intrinsically fluorescent photoclick HA nanogels (HA-NGs) designed for the highly efficient loading and active intracellular delivery of cytochrome C (CC) (Fig. 4), a powerful apoptotic protein, specifically targeting CD44 in xenografted breast tumors in mice (Li et al., 2016) (Fig. 6). HANGs are synthesized from hyaluronic acid-grafted oligo(ethylene glycol)-tetrazole (HA-OEG-Tet) through a process of inverse nanoprecipitation followed by photoclick cross-linking utilizing the "tetrazole-alkene" reaction with L-cystine dimethacrylamide (MA-Cys-MA).

Preactivated HA demonstrated a 3.6-fold prolonged disintegration time compared to unmodified HA as well as a prolonged mucoadhesion time, presumably leading to an increased contact time of the drug with the adsorption sites *in vivo.* Moreover, these modified HAs were found to be biocompatible and therefore they are safe tools for vaginal application.

Thermogels have been extensively studied over the past three due to their unique mechanisms and potential applications, such as drug delivery, three-cell culture,

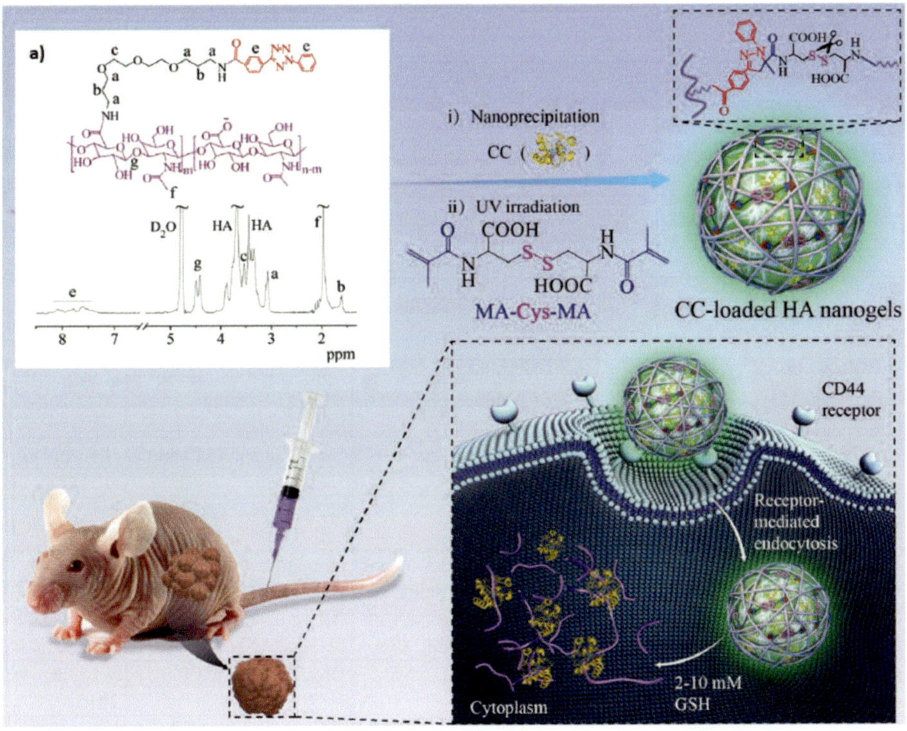

Fig. 6. An illustration of redox-sensitive and intrinsically fluorescent photoclick hyaluronic acid nanogels designed for the highly efficient encapsulation and targeted delivery of cytochrome c to breast tumors with (a) H NMR spectrum of NH2-OEG-Tet showed clear signals attributed to Tet (δ 8.67, 8.25, 8.17, 8.07, 7.71, and 7.64) and TTDA moieties (δ 3.52, 3.46, 3.42, 2.59, 1.79, and 1.57) (Li et al., 2021).

wound healing, imaging, prevention of post-surgical tissue adhesion, and ophthalmic applications (Cui et al., 2021; Cui et al., 2018, 2019; Kim et al., 2022; Liu et al., 2019). Hyaluronic acid-graft-poly(propylene glycol) (HA-g-PPG) was synthesized to induce hydrophobic interactions between HA-g-PPG F127 PPG (poly(ethylene glycol)-poly(propylene glycol)-poly(ethylene glycol)), in increased gel stability of F127 gel (Kim et al., 2022). More studies and information are highlighted in Table 1.

Table 1. Structure and application of redox sensitive hyaluronic acid based carriers.

Structure	Drug/Agent	Delivery system	Application/Cell type	Ref.
Zeolitic imidazolate framework-8/HA/polydopamine	Doxorubicin	Metal-organic framework	Prostate cancer/PC-3 cells	(Shu et al., 2018)
Folic acid-HA-SS-vitamin E succinate	Paclitaxel	Dual-targeted micelle	MCF-7 and NIH3T3 cells	(Yang et al., 2020)
HA, Chitosan	FITC-dextran	Hydrogel, Microcapsules (layer-by-layer self-assembled)	HEK 293 and HeLa cells	(Yang et al., 2018)
mPEG and hexadecanol-modified - HA	Paclitaxel	Micelle	MCF-7 and H22 cells/ H22 tumor-bearing mice	(Jiyang Liu et al., 2020)
HA-ss-tocopherol succinate	Paclitaxel	Micelle	Melanoma/B16F10 cells	(Xia et al., 2018)
HA-Cholesterol/ Dasatinib(cystamine dihydrochloride, and hexamethylene diamine as linkers)	Dasatinib-doxorubicin	Micelle	HeLa In a redox environment (10 mM dithiothreitol)	(Lin et al., 2021)
HA-Chlorin e6 Tetramer	Chlorin e6	Self-assembled nano -photosensitizers	B16F10 and CT26 cells	(Kim et al., 2019)
HA-coated complex	Epirubicin	Decorated niosome	Breast cancer/4T1 and SkBr3 cells	(Mansoori-Kermani et al., 2022)
HA-diselenide bond-carboxylic aid -poly (amidoamine)	siRNA	Hybrid nanocomplex	Lung carcinoma/A549 cells/A549 tumor-bearing mice	(Chen et al., 2022)
HA-functionalized nano-Graphene oxide	Gefitinib	Nanosheet	Lung cancer/A549 and HELF cells	(Jian Liu et al., 2018)
			Hela, HepG2 and 293T cell	
Biodegradable Thiol-Hyaluronic Acid/Chitosan	Doxorubicin	Dual cross-linked nanoparticles	Human breast cancer cells/SKBR3	(Xia et al., 2021)
HA-cys-DHA/Ce6	Docetaxel	Nanoparticle/self-assembled	Breast cancer/MCF-7	(Wang et al., 2021)
Fluorescent photoclick HA	Cytochrome c	Nanogel	Breast cancer/MCF-7, U87	(Li et al., 2016)
HA/Aminoethyl disulfide	-	Hydrogel	Chronic wound healing/ L929	(Gao et al., 2019)
HA, RHAMM or CD168	Doxorubicin	Nanogel	LNCaP, H22, A549, NIH3T3	(Yang et al., 2017)
Gold nanorods/cystamine linker/HA	Doxorubicin	Nanogel	Breast cancer/MCF-7	(Li et al., 2019)

6. Conclusion

Numerous research efforts have concentrated on exploiting HA's ability to target tumors for the delivery of anticancer medications. Redox-sensitive modification of HAs, owing to their multifunctional properties, can be employed across a broad spectrum of pharmaceutical applications. Due to their notable characteristics, redox-sensitive HA carriers find application in cancer therapy. To summarize, creating innovative HA production systems emerges as a crucial objective for the near future, to address the growing demand for novel modifications of HA at affordable prices. Simultaneously, the discovery of new HA bioactive mechanisms, coupled with a comprehensive understanding of HA product effectiveness, continues to drive and inspire advancements in biomedical and pharmaceutical research.

References

Akita, H., Ishiba, R., Hatakeyama, H., Tanaka, H., Sato, Y., Tange, K. et al. 2013. A neutral envelope-type nanoparticle containing pH-responsive and SS-cleavable lipid-like material as a carrier for plasmid DNA. Advanced Healthcare Materials 2(8): 1120–1125.

Allawadhi, P., Singh, V., Govindaraj, K., Khurana, I., Sarode, L.P., Navik, U. et al. 2022. Biomedical applications of polysaccharide nanoparticles for chronic inflammatory disorders: Focus on rheumatoid arthritis, diabetes and organ fibrosis. Carbohydrate Polymers 281: 118923.

Amorim, S., Reis, C.A., Reis, R.L. and Pires, R.A. 2021. Extracellular matrix mimics using hyaluronan-based biomaterials. Trends in Biotechnology 39(1): 90–104.

Arnér, E.S. and Holmgren, A. 2000. Physiological functions of thioredoxin and thioredoxin reductase. European Journal of Biochemistry 267(20): 6102–6109.

Aruffo, A., Stamenkovic, I., Melnick, M., Underhill, C.B. and Seed, B. 1990. CD44 is the principal cell surface receptor for hyaluronate. Cell 61(7): 1303–1313.

Bernkop-Schnürch, A., Hornof, M. and Zoidl, T. 2003. Thiolated polymers—thiomers: synthesis and *in vitro* evaluation of chitosan–2-iminothiolane conjugates. International Journal of Pharmaceutics 260(2): 229–237.

Brown, T., Laurent, U. and JR, F. 1991. Turnover of hyaluronan in synovial joints: elimination of labelled hyaluronan from the knee joint of the rabbit. Experimental Physiology: Translation and Integration 76(1): 125–134.

Cao, W., Sui, J., Ma, M., Xu, Y., Lin, W., Chen, Y. et al. 2019. The preparation and biocompatible evaluation of injectable dual crosslinking hyaluronic acid hydrogels as cytoprotective agents. Journal of Materials Chemistry B 7(28): 4413–4423.

Censi, R., Fieten, P.J., di Martino, P., Hennink, W.E. and Vermonden, T. 2010. *In situ* forming hydrogels by tandem thermal gelling and Michael addition reaction between thermosensitive triblock copolymers and thiolated hyaluronan. Macromolecules 43(13): 5771–5778.

Chen, D., Zhang, P., Li, M., Li, C., Lu, X., Sun, Y. et al. 2022. Hyaluronic acid-modified redox-sensitive hybrid nanocomplex loading with siRNA for non-small-cell lung carcinoma therapy. Drug Delivery 29(1): 574–587.

Chen, M., Liu, D., Liu, F., Wu, Y., Peng, X. and Song, F. 2021. Recent advances of redox-responsive nanoplatforms for tumor theranostics. Journal of Controlled Release 332: 269–284.

Cheng, G., He, Y., Xie, L., Nie, Y., He, B., Zhang, Z. et al. 2012. Development of a reduction-sensitive diselenide-conjugated oligoethylenimine nanoparticulate system as a gene carrier. International Journal of Nanomedicine 3991–4006.

Cheng, R., Meng, F., Deng, C. and Zhong, Z. 2015. Bioresponsive polymeric nanotherapeutics for targeted cancer chemotherapy. Nano Today 10(5): 656–670.

Chu, X., Tian, W., Ning, J., Xiao, G., Zhou, Y., Wang, Z. et al. 2024. Cancer stem cells: advances in knowledge and implications for cancer therapy. Signal Transduction and Targeted Therapy 9(1): 170.

Clausen, A.E., Kast, C.E. and Bernkop-Schnürch, A. 2002. The role of glutathione in the permeation enhancing effect of thiolated polymers. Pharmaceutical Research 19: 602–608.

Cui, S., Wei, Y., Bian, Q., Zhu, Y., Chen, X., Zhuang, Y. et al. 2021. Injectable thermogel generated by the "block blend" strategy as a biomaterial for endoscopic submucosal dissection. ACS Applied Materials & Interfaces 13(17): 19778–19792.

Cui, S., Yu, L. and Ding, J. 2018. Semi-bald micelles and corresponding percolated micelle networks of thermogels. Macromolecules 51(16): 6405–6420.

Cui, S., Yu, L. and Ding, J. 2019. Thermogelling of amphiphilic block copolymers in water: ABA type versus AB or BAB type. Macromolecules 52(10): 3697–3715.

Dan, Z., Cao, H., He, X., Zeng, L., Zou, L., Shen, Q. et al. 2015. Biological stimuli-responsive cyclodextrin-based host–guest nanosystems for cancer therapy. International Journal of Pharmaceutics 483(1-2): 63–68.

de Sousa, I.P., Suchaoin, W., Zupančič, O., Leichner, C. and Bernkop-Schnürch, A. 2016. Totally S-protected hyaluronic acid: Evaluation of stability and mucoadhesive properties as liquid dosage form. Carbohydrate Polymers 152: 632–638.

DM, T. 2003. The importance of glutathione in human disease. Biomed Pharmacother. 57: 145–155.

Eslami, M., Memarsadeghi, O., Davarpanah, A., Arti, A., Nayernia, K. and Behnam, B. 2024. Overcoming chemotherapy resistance in metastatic cancer: A comprehensive review. Biomedicines 12(1): 183.

Fu, C., Li, H., Li, N., Miao, X., Xie, M., Du, W. et al. 2015. Conjugating an anticancer drug onto thiolated hyaluronic acid by acid liable hydrazone linkage for its gelation and dual stimuli-response release. Carbohydrate Polymers 128: 163–170.

Fu, L.H., Wan, Y., Qi, C., He, J., Li, C., Yang, C. et al. 2021. Nanocatalytic theranostics with glutathione depletion and enhanced reactive oxygen species generation for efficient cancer therapy. Advanced Materials 33(7): 2006892.

Gallo, N., Nasser, H., Salvatore, L., Natali, M.L., Campa, L., Mahmoud, M. et al. 2019. Hyaluronic acid for advanced therapies: Promises and challenges. European Polymer Journal 117: 134–147.

Gao, Z., Golland, B., Tronci, G. and Thornton, P.D. 2019. A redox-responsive hyaluronic acid-based hydrogel for chronic wound management. Journal of Materials Chemistry B 7(47): 7494–7501.

Gilbert, H.F. 1995. [2] Thiol/disulfide exchange equilibria and disulfidebond stability. Methods in Enzymology 251: 8–28.

Go, Y.-M. and Jones, D.P. 2008. Redox compartmentalization in eukaryotic cells. Biochimica et Biophysica Acta (BBA)-General Subjects 1780(11): 1273–1290.

Griesser, J., Hetényi, G. and Bernkop-Schnürch, A. 2018. Thiolated hyaluronic acid as versatile Mucoadhesive polymer: from the chemistry behind to product developments—what are the capabilities? Polymers 10(3): 243.

Gupta, C., Naik, I., Menon, M., Ambre, P. and Coutinho, E. 2023. A review on exploring the opportunities of polymer drug conjugated systems for targeted cancer treatment. Current Drug Delivery 20(1): 8–30.

Hari, S.K., Gauba, A., Shrivastava, N., Tripathi, R.M., Jain, S.K. and Pandey, A.K. 2023. Polymeric micelles and cancer therapy: An ingenious multimodal tumor-targeted drug delivery system. Drug Delivery and Translational Research 13(1): 135–163.

Hossen, S., Hossain, M.K., Basher, M., Mia, M., Rahman, M. and Uddin, M.J. 2019. Smart nanocarrier-based drug delivery systems for cancer therapy and toxicity studies: A review. Journal of Advanced Research 15: 1–18.

Hou, J., Zhao, Y., Sun, L. and Zou, X. 2023. Enzyme/GSH/pH-responsive hyaluronic acid grafted porous silica nanocarriers bearing Ag2S QDs for fluorescence imaging and combined therapy. Carbohydrate Polymers 305: 120547.

Huang, Y.-Q., Zhang, R., Zhao, Y.-K., Chen, H., Jiang, R.-C., Liu, X.-F. et al. 2017. Self-assembled nanoparticles based on a cationic conjugated polymer/hyaluronan–cisplatin complex as a multifunctional platform for simultaneous tumor-targeting cell imaging and drug delivery. New Journal of Chemistry 41(12): 4998–5006.

Huang, Y., Yao, X., Zhang, R., Ouyang, L., Jiang, R., Liu, X. et al. 2014. Cationic conjugated polymer/fluoresceinamine-hyaluronan complex for sensitive fluorescence detection of CD44 and tumor-targeted cell imaging. ACS Applied Materials & Interfaces 6(21): 19144–19153.

Iqbal, J., Shahnaz, G., Dünnhaupt, S., Müller, C., Hintzen, F. and Bernkop-Schnürch, A. 2012. Preactivated thiomers as mucoadhesive polymers for drug delivery. Biomaterials 33(5): 1528–1535.

Jiang, Y., Chen, J., Deng, C., Suuronen, E.J. and Zhong, Z. 2014. Click hydrogels, microgels and nanogels: Emerging platforms for drug delivery and tissue engineering. Biomaterials 35(18): 4969–4985.

Kafedjiiski, K., Jetti, R.K., Föger, F., Hoyer, H., Werle, M., Hoffer, M. et al. 2007. Synthesis and *in vitro* evaluation of thiolated hyaluronic acid for mucoadhesive drug delivery. International Journal of Pharmaceutics 343(1-2): 48–58.

Kieliszek, M. and Lipinski, B. 2020. Selenium supplementation in the prevention of coronavirus infections (COVID-19). Medical Hypotheses 143: 109878.

Kim, D.-M., Shim, Y.H., Kwon, H., Kim, J.-P., Park, J.-I., Kim, D.H. et al. 2019. CD44 receptor–specific and redox-sensitive nanophotosensitizers of hyaluronic acid–chlorin E6 tetramer having diselenide linkages for photodynamic treatment of cancer cells. Journal of Pharmaceutical Sciences 108(11): 3713–3722.

Kim, S., Lee, H.J. and Jeong, B. 2022. Hyaluronic acid-g-PPG and PEG-PPG-PEG hybrid thermogel for prolonged gel stability and sustained drug release. Carbohydrate Polymers 291: 119559.

Kotla, N.G., Bonam, S.R., Rasala, S., Wankar, J., Bohara, R.A., Bayry, J. et al. 2021. Recent advances and prospects of hyaluronan as a multifunctional therapeutic system. Journal of Controlled Release 336: 598–620.

Kotla, N.G., Mohd Isa, I.L., Larrañaga, A., Maddiboyina, B., Swamy, S.K., Sivaraman, G. et al. 2023. Hyaluronic acid-based bioconjugate systems, scaffolds, and their therapeutic potential. Advanced Healthcare Materials 12(20): 2203104.

Kou, L., Jiang, X., Huang, H., Lin, X., Zhang, Y., Yao, Q. et al. 2020. The role of transporters in cancer redox homeostasis and cross-talk with nanomedicines. Asian Journal of Pharmaceutical Sciences 15(2): 145–157.

Köwitsch, A., Niepel, M.S., Michanetzis, G.P., Missirlis, Y.F. and Groth, T. 2016. Effect of immobilized thiolated glycosaminoglycans on fibronectin adsorption and behavior of fibroblasts. Macromolecular Bioscience 16(3): 381–394.

Laffleur, F., Röggla, J., Idrees, M.A. and Griessinger, J. 2014. Chemical modification of hyaluronic acid for intraoral application. Journal of Pharmaceutical Sciences 103(8): 2414–2423.

Laffleur, F., Psenner, J. and Suchaoin, W. 2015a. Permeation enhancement via thiolation: *in vitro* and *ex vivo* evaluation of hyaluronic acid-cysteine ethyl ester. Journal of Pharmaceutical Sciences 104(7): 2153–2160.

Laffleur, F., Wagner, J. and Barthelmes, J. 2015b. A potential tailor-made hyaluronic acid buccal delivery system comprising rotigotine for Parkinson's disease? Future Medicinal Chemistry 7(10): 1225–1232.

Laffleur, F., Schmelzle, F., Ganner, A. and Vanicek, S. 2017. *In vitro* and *ex vivo* evaluation of novel curcumin-loaded excipient for buccal delivery. AAPS PharmSciTech 18: 2102–2109.

Lallana, E., Rios de la Rosa, J.M., Tirella, A., Pelliccia, M., Gennari, A., Stratford, I.J. et al. 2017. Chitosan/hyaluronic acid nanoparticles: rational design revisited for RNA delivery. Molecular Pharmaceutics 14(7): 2422–2436.

Laurent, T.C. and Fraser, J.R.E. 1992. Hyaluronan 1. The FASEB Journal 6(7): 2397–2404.

Lee, S.Y., Kang, M.S., Jeong, W.Y., Han, D.-W. and Kim, K.S. 2020. Hyaluronic acid-based theranostic nanomedicines for targeted cancer therapy. Cancers 12(4): 940.

Li, B., Xu, Q., Li, X., Zhang, P., Zhao, X. and Wang, Y. 2019. Redox-responsive hyaluronic acid nanogels for hyperthermia-assisted chemotherapy to overcome multidrug resistance. Carbohydrate Polymers 203: 378–385.

Li, H., Zhuang, S., Yang, Y., Zhou, F., Rong, J. and Zhao, J. 2021. ATP/Hyals dually responsive core-shell hyaluronan/chitosan-based drug nanocarrier for potential application in breast cancer therapy. International Journal of Biological Macromolecules 183: 839–851.

Li, J., Huo, M., Wang, J., Zhou, J., Mohammad, J.M., Zhang, Y. et al. 2012. Redox-sensitive micelles self-assembled from amphiphilic hyaluronic acid-deoxycholic acid conjugates for targeted intracellular delivery of paclitaxel. Biomaterials 33(7): 2310–2320.

Li, J., Yin, T., Wang, L., Yin, L., Zhou, J. and Huo, M. 2015. Biological evaluation of redox-sensitive micelles based on hyaluronic acid-deoxycholic acid conjugates for tumor-specific delivery of paclitaxel. International Journal of Pharmaceutics 483(1-2): 38–48.

Li, S., Zhang, J., Deng, C., Meng, F., Yu, L. and Zhong, Z. 2016. Redox-sensitive and intrinsically fluorescent photoclick hyaluronic acid nanogels for traceable and targeted delivery of cytochrome c to breast tumor in mice. ACS Applied Materials & Interfaces 8(33): 21155–21162.

Li, W., Li, M. and Qi, J. 2021. Nano-drug design based on the physiological properties of glutathione. Molecules 26(18): 5567.

Li, X., Yu, G., Jin, K. and Yin, Z. 2012. Hyaluronic acid L-cysteine conjugate exhibits controlled-release potential for mucoadhesive drug delivery. Die Pharmazie-An International Journal of Pharmaceutical Sciences 67(3): 224–228.

Li, Y., Maciel, D., Rodrigues, J., Shi, X. and Tomás, H. 2015. Biodegradable polymer nanogels for drug/nucleic acid delivery. Chemical Reviews 115(16): 8564–8608.

Liang, K., Ng, S., Lee, F., Lim, J., Chung, J.E., Lee, S.S. et al. 2016. Targeted intracellular protein delivery based on hyaluronic acid–green tea catechin nanogels. Acta Biomaterialia 33: 142–152.

Lin, Y.-K., Wang, S.-W. and Lee, R.-S. 2021. Redox-responsive dasatinib-containing hyaluronic acid prodrug and co-delivery of doxorubicin for cancer therapy. International Journal of Polymeric Materials and Polymeric Biomaterials 70(18): 1329–1343.

Liu, J., Liang, N., Li, S., Han, Y., Yan, P., Kawashima, Y. et al. 2020. Tumor-targeting and redox-sensitive micelles based on hyaluronic acid conjugate for delivery of paclitaxel. Journal of Biomaterials Applications 34(10): 1458–1469.

Liu, J., Zhang, D., Lian, S., Zheng, J., Li, B., Li, T. et al. 2018. Redox-responsive hyaluronic acid-functionalized graphene oxide nanosheets for targeted delivery of water-insoluble cancer drugs. International Journal of Nanomedicine 7457–7472.

Liu, Y., Shu, X.Z. and Prestwich, G.D. 2005. Biocompatibility and stability of disulfide-crosslinked hyaluronan films. Biomaterials 26(23): 4737–4746.

Liu, Z., Liow, S.S., Lai, S.L., Alli-Shaik, A., Holder, G.E., Parikh, B.H. et al. 2019. Retinal-detachment repair and vitreous-like-body reformation via a thermogelling polymer endotamponade. Nature Biomedical Engineering 3(8): 598–610.

Lu, Y., Sun, W. and Gu, Z. 2014. Stimuli-responsive nanomaterials for therapeutic protein delivery. Journal of Controlled Release 194: 1–19.

Luo, Y. and Prestwich, G.D. 1999. Synthesis and selective cytotoxicity of a hyaluronic acid− antitumor bioconjugate. Bioconjugate Chemistry 10(5): 755–763.

Luo, Y., Ziebell, M.R. and Prestwich, G.D. 2000. A hyaluronic acid− taxol antitumor bioconjugate targeted to cancer cells. Biomacromolecules 1(2): 208–218.

Lv, S., Tang, Z., Zhang, D., Song, W., Li, M., Lin, J. et al. 2014. Well-defined polymer-drug conjugate engineered with redox and pH-sensitive release mechanism for efficient delivery of paclitaxel. Journal of Controlled Release 194: 220–227.

Ma, N., Li, Y., Xu, H., Wang, Z. and Zhang, X. 2010. Dual redox responsive assemblies formed from diselenide block copolymers. Journal of the American Chemical Society 132(2): 442–443.

Mahdieh, A., Motasadizadeh, H., Yeganeh, H., Nyström, B. and Dinarvand, R. 2022. Redox-responsive waterborne polyurethane nanocarriers for targeted doxorubicin delivery. International Journal of Pharmaceutics 628: 122275.

Mansoori-Kermani, A., Khalighi, S., Akbarzadeh, I., Niavol, F.R., Motasadizadeh, H., Mahdieh, A. et al. 2022. Engineered hyaluronic acid-decorated niosomal nanoparticles for controlled and targeted delivery of epirubicin to treat breast cancer. Materials Today Bio 16: 100349.

Meng, F., Hennink, W.E. and Zhong, Z. 2009. Reduction-sensitive polymers and bioconjugates for biomedical applications. Biomaterials 30(12): 2180–2198.

Meng, X., Shen, Y., Zhao, H., Lu, X., Wang, Z. and Zhao, Y. 2024. Redox-manipulating nanocarriers for anticancer drug delivery: a systematic review. Journal of Nanobiotechnology 22(1): 587.

Meyer, K. and Palmer, J.W. 1934. The polysaccharide of the vitreous humor. Journal of Biological Chemistry 107(3): 629–634.

Oh, J.K., Drumright, R., Siegwart, D.J. and Matyjaszewski, K. 2008. The development of microgels/nanogels for drug delivery applications. Progress in Polymer Science 33(4): 448–477.

Oh, S., Wilcox, M., Pearson, J.P. and Borrós, S. 2015. Optimal design for studying mucoadhesive polymers interaction with gastric mucin using a quartz crystal microbalance with dissipation (QCM-D): Comparison of two different mucin origins. European Journal of Pharmaceutics and Biopharmaceutics 96: 477–483.

Peng, H.-H., Chen, Y.-M., Lee, C.-I. and Lee, M.-W. 2013. Synthesis of a disulfide cross-linked polygalacturonic acid hydrogel for biomedical applications. Journal of Materials Science: Materials in Medicine 24: 1375–1382.
Prestwich, G.D. 2011. Hyaluronic acid-based clinical biomaterials derived for cell and molecule delivery in regenerative medicine. Journal of Controlled Release 155(2): 193–199.
Raina, S. and Missiakas, D. 1997. Making and breaking disulfide bonds. Annual Review of Microbiology 51(1): 179–202.
Raza, A., Rasheed, T., Nabeel, F., Hayat, U., Bilal, M. and Iqbal, H.M. 2019. Endogenous and exogenous stimuli-responsive drug delivery systems for programmed site-specific release. Molecules 24(6): 1117.
Savic, R., Luo, L., Eisenberg, A. and Maysinger, D. 2003. Micellar nanocontainers distribute to defined cytoplasmic organelles. Science 300(5619): 615–618.
Serban, M.A., Yang, G. and Prestwich, G.D. 2008. Synthesis, characterization and chondroprotective properties of a hyaluronan thioethyl ether derivative. Biomaterials 29(10): 1388–1399.
Shinkar, D.M., Dhake, A.S. and Setty, C.M. 2012. Drug delivery from the oral cavity: A focus on mucoadhesive. PDA J. Pharm. Sci. Technol. 66: 466–500.
Shu, F., Lv, D., Song, X.-L., Huang, B., Wang, C., Yu, Y. et al. 2018. Fabrication of a hyaluronic acid conjugated metal organic framework for targeted drug delivery and magnetic resonance imaging. RSC Advances 8(12): 6581–6589.
Song, M., Liang, Y., Li, K., Zhang, J., Zhang, N., Tian, B. et al. 2019. Hyaluronic acid modified liposomes for targeted delivery of doxorubicin and paclitaxel to CD44 overexpressing tumor cells with improved dual-drugs synergistic effect. Journal of Drug Delivery Science and Technology 53: 101179.
Stern, R., Kogan, G., Jedrzejas, M.J. and Šoltés, L. 2007. The many ways to cleave hyaluronan. Biotechnology Advances 25(6): 537–557.
Sun, Y.-X., Xiao, W., Cheng, S.-X., Zhang, X.-Z. and Zhuo, R.-X. 2008. Synthesis of (Dex-HMDI)-g-PEIs as effective and low cytotoxic nonviral gene vectors. Journal of Controlled Release 128(2): 171–178.
Tahara, Y. and Akiyoshi, K. 2015. Current advances in self-assembled nanogel delivery systems for immunotherapy. Advanced Drug Delivery Reviews 95: 65–76.
Toole, B.P. 2004. Hyaluronan: from extracellular glue to pericellular cue. Nature Reviews Cancer 4(7): 528–539.
Tsarevsky, N.V. and Matyjaszewski, K. 2002. Reversible redox cleavage/coupling of polystyrene with disulfide or thiol groups prepared by atom transfer radical polymerization. Macromolecules 35(24): 9009–9014.
Tuffaha, H., Edmunds, K., Fairbairn, D., Roberts, M.J., Chambers, S., Smith, D.P. et al. 2024. Guidelines for genetic testing in prostate cancer: a scoping review. Prostate Cancer and Prostatic Diseases 27(4): 594–603.
Turcsányi, Á., Varga, N. and Csapó, E. 2020. Chitosan-modified hyaluronic acid-based nanosized drug carriers. International Journal of Biological Macromolecules 148: 218–225.
Valko, M., Rhodes, C., Moncol, J., Izakovic, M. and Mazur, M. 2006. Free radicals, metals and antioxidants in oxidative stress-induced cancer. Chemico-biological Interactions 160(1): 1–40.
Wang, R., Yang, H., Khan, A.R., Yang, X., Xu, J., Ji, J. et al. 2021. Redox-responsive hyaluronic acid-based nanoparticles for targeted photodynamic therapy/chemotherapy against breast cancer. Journal of Colloid and Interface Science 598: 213–228.
Wen, H., Dong, C., Dong, H., Shen, A., Xia, W., Cai, X. et al. 2012. Engineered redox-responsive PEG detachment mechanism in PEGylated nano-graphene oxide for intracellular drug delivery. Small 8(5): 760–769.
Xia, D., Wang, F., Pan, S., Yuan, S., Liu, Y. and Xu, Y. 2021. Redox/ph-responsive biodegradable thiol-hyaluronic acid/chitosan charge-reversal nanocarriers for triggered drug release. Polymers 13(21): 3785.
Xia, J., Du, Y., Huang, L., Chaurasiya, B., Tu, J., Webster, T.J. et al. 2018. Redox-responsive micelles from disulfide bond-bridged hyaluronic acid-tocopherol succinate for the treatment of melanoma. Nanomedicine: Nanotechnology, Biology and Medicine 14(3): 713–723.

Yang, C., Li, C., Zhang, P., Wu, W. and Jiang, X. 2017. Redox responsive hyaluronic acid nanogels for treating RHAMM (CD168) over-expressive cancer, both primary and metastatic tumors. Theranostics 7(6): 1719.

Yang, H., Wang, N., Mo, L., Wu, M., Yang, R., Xu, X. et al. 2019. Reduction sensitive hyaluronan-SS-poly (ε-caprolactone) block copolymers as theranostic nanocarriers for tumor diagnosis and treatment. Materials Science and Engineering: C 98: 9–18.

Yang, Y., Li, Y., Chen, K., Zhang, L., Qiao, S., Tan, G. et al. 2020. Dual receptor-targeted and redox-sensitive polymeric micelles self-assembled from a folic acid-hyaluronic acid-SS-vitamin E succinate polymer for precise cancer therapy. International Journal of Nanomedicine 2885–2902.

Yang, Y., Zhu, H., Wang, J., Fang, Q. and Peng, Z. 2018. Enzymatically disulfide-crosslinked chitosan/hyaluronic acid layer-by-layer self-assembled microcapsules for redox-responsive controlled release of protein. ACS Applied Materials & Interfaces 10(39): 33493–33506.

Yin, S., Huai, J., Chen, X., Yang, Y., Zhang, X., Gan, Y. et al. 2015. Intracellular delivery and antitumor effects of a redox-responsive polymeric paclitaxel conjugate based on hyaluronic acid. Acta Biomaterialia 26: 274–285.

Yin, T., Wang, J., Yin, L., Shen, L., Zhou, J. and Huo, M. 2015. Redox-sensitive hyaluronic acid–paclitaxel conjugate micelles with high physical drug loading for efficient tumor therapy. Polymer Chemistry 6(46): 8047–8059.

Yu, M.K., Park, J. and Jon, S. 2012. Targeting strategies for multifunctional nanoparticles in cancer imaging and therapy. Theranostics 2(1): 3.

Zhang, L., Gao, S., Zhang, F., Yang, K., Ma, Q. and Zhu, L. 2014. Activatable hyaluronic acid nanoparticle as a theranostic agent for optical/photoacoustic image-guided photothermal therapy. ACS Nano 8(12): 12250–12258.

Zhang, X., Malhotra, S., Molina, M. and Haag, R. 2015. Micro-and nanogels with labile crosslinks–from synthesis to biomedical applications. Chemical Society Reviews 44(7): 1948–1973.

Zhang, Z.-T., Huang-Fu, M.-Y., Xu, W.-H. and Han, M. 2019. Stimulus-responsive nanoscale delivery systems triggered by the enzymes in the tumor microenvironment. European Journal of Pharmaceutics and Biopharmaceutics 137: 122–130.

Chapter 9

Catechol-Based Redox Polymers for Energy Storage Devices

Fatemeh Mahdavi,[1,2,3] *Helma Vakili,*[4]
Seyed Mohammad Amin Hashemi[5,*] and *Atefeh Derakhshani*[3,6,*]

1. Introduction

In the realm of electrochemistry, three primary classes of polymers are prevalent: ion-exchange polymers, conductive polymers, and redox polymers, Fig. 1. Ion-exchange polymers (Fig. 1, A), characterized by a non-conducting backbone bearing ionic pendant groups, find utility in sensors for imparting ionic selectivity and in energy conversion/storage devices such as fuel cells and electrolyzers for facilitating ionic transport. Conductive polymers, Fig. 1, B, distinguished by their intrinsically conductive backbone, have been employed to enhance the selectivity and sensitivity of electrochemical sensors and as functional additives in supercapacitors. The third prominent category, redox polymers, crucial in electrochemical devices,

[1] Department of Medical Nanotechnology, School of Advanced Technologies in Medicine, Tehran University of Medical Sciences, Tehran, Iran.
[2] Research Center for Nuclear Medicine, Shariati Hospital, Tehran University of Medical Sciences, Tehran, Iran.
[3] Students' Scientific Research Center, Tehran University of Medical Sciences, Tehran, Iran.
[4] Polymer Engineering group, School of Chemical Engineering, College of Engineering, University of Tehran, Tehran, Iran.
[5] Department of Tissue Engineering and Applied Cell Sciences, School of Advanced Technologies in Medicine, Shahid Beheshti University of Medical Sciences, Tehran, Iran.
[6] Department of Tissue Engineering, School of Advanced Technologies in Medicine, Tehran University of Medical Sciences, Tehran, Iran.
* Corresponding author: smah77.maph@gmail.com; atefe.drkhshani93@gmail.com
DOI: 10.1201/9781003546672-9

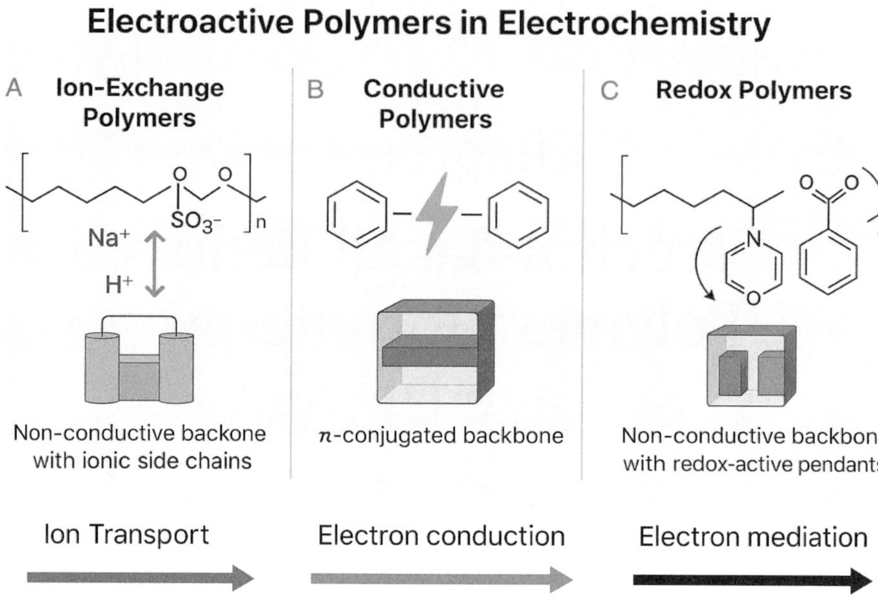

Fig. 1. Schematic of primary classes of polymers in electrochemistry.

comprises a non-conductive backbone functionalized with pendant chains incorporating redox-active moieties, Fig. 1, C. These polymeric systems can mediate electron transport through a mechanism based on self-exchange conductivity. The facile modulation between oxidation and reduction states at the redox-active sites, achieved through the reversible gain or loss of electrons, renders redox polymers highly suitable materials for energy conversion and storage technologies, including redox flow batteries and supercapacitors (Darmayanti et al., 2024).

Redox polymers have attracted significant interest as electroactive components for electrochemical energy storage (EES) systems. The inorganic materials presently employed in these applications impose a substantial burden on the raw material supply chain and pose environmental concerns stemming from their limited availability, inherent toxicity (e.g., Co, V, Ni), and challenges in recycling. Occasionally, these metallic substances encounter challenges regarding harmful effects and ecological concerns associated with waste and processes that utilize extraction techniques which emit hazardous materials (Tchounwou et al., 2012). Moreover, these metallic materials rely on variations in the metal's oxidation state for charge retention, along with an accompanying equilibrium of the charged arrangement using particular counterions. In contrast, organic compounds are constituted by naturally abundant elements (C, H, N, O, S) and, with judicious design, can even be synthesized from renewable feedstocks. The advancement of redox-active polymers as organic electrode materials is envisioned as a viable answer for secure, sustainable, and efficient electrochemical energy storage technologies. Their redox characteristics can be readily tailored through modifications of the chemical constitution and macromolecular architecture. In every redox process, as demonstrated in Fig. 2, one

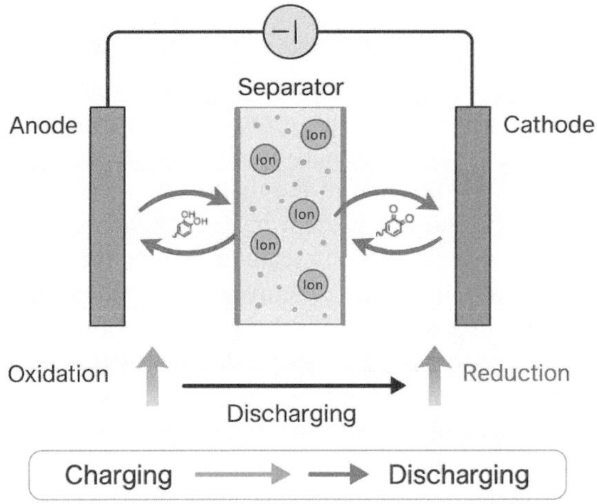

Fig. 2. Diagram of the process of reduction-oxidation.

material undergoes oxidation while another experiences reduction at the same time. These reactions are essential to energy transformation activities, such as batteries, rusting, burning, and metabolic functions. Electroactive macromolecules feature discrete sites or functional groups capable of undergoing oxidation and reduction processes. The electronic, optical, mechanical, or chemical attributes of redox polymers are contingent upon their prevailing oxidation state. Consequently, these polymeric materials are finding burgeoning applications in materials science and are increasingly being integrated into the design of diverse electrochemical devices (Ma et al., 2023).

Moreover, electroactive macromolecules function as charge-conducting channels, interfacing the electrode with biocatalysts (enzymes), subcellular compartments (organelles), or microbial entities (bacteria), thereby markedly augmenting the operational efficacy of bioelectrosynthetic platforms, bioanodic fuel cells, or biological sensors. Along with the clearer understanding of electron translocation pathways within biological systems, the judicious engineering of redox polymers has witnessed increasing scholarly interest. Furthermore, organic moieties and redox-active polymers can manifest elevated theoretical gravimetric capacities, reaching magnitudes up to 600 mAh g^{-1}, in contradistinction to inorganic cathodic materials (typically < 250 mAh g^{-1}), thereby positioning them as compelling candidates for advanced-generation electrochemical accumulators and energy storage capacitors. Notwithstanding these attributes, extant performance constraints concerning operational potential, electrochemical durability (cyclability), intrinsic charge leakage (self-discharge), and overall device robustness continue to limit their widespread technological deployment (Darmayanti et al., 2024).

Adhesive proteins extracted from mussels, featuring catecholic amino acids such as 3, 4-dihydroxy-L-phenylalanine, serve as valuable resources for biomedical, industrial, and pharmaceutical uses due to their unique binding capabilities, hydrophilic nature, permeability, and resilience on diverse surfaces in both air and water environments (Glass et al., 2009; Silverman and Roberto, 2007; Yu and Deming, 1998). Notably, the groundbreaking identification of adhesive proteins derived from catechol, which facilitate strong attachment of invertebrate blue mussels to nearly any surface, conducted by Waite et al. in the early 1980s, has resulted in a significant breakthrough in the realm of materials science (Waite and Tanzer, 1981). The catechol groups present in 3, 4-dihydroxy-L-phenylalanine are widely recognized for their capability to create numerous forms of chemical bonds, such as cross-linking through oxidation, consequently demonstrating remarkable adherence qualities for a diverse range of substances, including polymers, inorganic metals, metal oxides, alloys, glass, ceramics, and wood (Saito and Yabu, 2015; Sedó et al., 2013b). Creating secure, affordable, practical, and efficient rechargeable batteries is essential for achieving a sustainable energy-driven, low-carbon footprint society. Recent findings have underscored that catechols present significantly more potential in various domains beyond adhesives and functional coatings, particularly in scenarios where they function as reversible redox components. Catechols hold significant value as cathode substances for energy storage owing to their impressive capacity and oxidation-reduction potential. Polymers tailored with catechols are now gaining significant attention in the development of the 'next-generation' safe, efficient, and eco-friendly electrochemical energy storage systems (Patil et al., 2018).

Recent investigations into catechol-derived redox polymers have aimed at enhancing their electrochemical efficiency and durability for energy storage applications. The polymer-based characteristics of these substances reduce solubility in electrolytic solutions, a frequent challenge associated with small organic compounds. This durability results in extended cycling longevity and reliable performance. A research investigation examined the macromolecular design of poly (catechol) electrodes to boost their performance in water-based zinc-polymer batteries, showing enhanced charge movement and overall capacity (Patil et al., 2021). A distinct study explored the application of atmospheric plasma deposition as a technique for fabricating durable redox-active thin films, emphasizing their prospects for use in lithium-ion battery technologies (Aqil et al., 2021). Moreover, progress in biopolymers sourced from catechol has resulted in innovative synthetic methods that enhance their structure and performance, rendering them increasingly suitable for eco-friendly energy storage (Patil et al., 2018). These investigations together advance the creation of future energy storage solutions that are more secure, highly effective, and eco-conscious. Polycatechols stand out as prominent examples of redox polymers utilized for energy storage, thanks to their impressive energy density and oxidation capabilities. A variety of synthetic (Lukyanov et al., 2019; Patil et al., 2017; Patil et al., 2018) and nature-inspired polycatechols (Lee et al., 2017; Patil et al., 2017; Sun et al., 2016) were evaluated as cathode components for various kinds of rechargeable batteries. This chapter aims to focus on redox polymers and catechol-based redox polymers from the viewpoint of energy storage applications due to their high theoretical capacity, reversible redox activity, and structural versatility. Particular attention is also devoted to their fabrication techniques, which greatly affect their electrochemical characteristics, durability, and manufacturability.

2. Redox Materials

While the energy sector currently represents the most prominent application of redox polymers, medicine is anticipated to become their most critical area of utilization in the coming decades. Historically, the primary application of redox polymers has been in electrochemical biosensors for analytes such as glucose and other bioactive molecules. However, a significant demand is emerging for redox polymers in cutting-edge healthcare technologies that are of paramount importance in contemporary medicine that we briefly discussed here.

These materials typically undergo chemical transformations in the presence and varying concentrations of oxidants and reductants. Considering the intricate nature of biological systems, it is paramount to ascertain whether the chemical response is influenced not only by the redox potential of the stimulus but also by its specific chemical characteristics. Moreover, it is essential to determine if distinct chemical responses can dictate divergent overall material functions. The innate and adaptive immune systems are indispensable for targeted pathogen defense and the establishment of immunological memory. Furthermore, the immune system plays a critical role in tissue repair. However, dysregulation of the immune system, either through insufficient or chronic activation, can contribute to a spectrum of diseases, including autoimmunity, cardiovascular disorders, and the acceleration of the physiological aging process.

Increased levels of reactive oxygen species (ROS) in immune cells may cause excessive inflammation, resulting in harm to tissues and the onset of diseases. Their function as intermediaries in biological processes positions them as significant players in a range of pathological conditions (Yeo et al., 2021).

ROS, including superoxide, hydrogen peroxide, and hydroxyl radicals, are products resulting from aerobic metabolism and exhibit a high degree of reactivity toward biological molecules. While they have been classically linked to oxidative stress and cellular harm, ROS also play crucial roles as signaling molecules within various physiological processes. From an evolutionary perspective, ROS may have developed as essential regulators for cellular adaptation to shifts in environmental conditions. In prokaryotic organisms, as shown in Fig. 3, A, ROS can directly stimulate transcription factors involved in the stress response. The study of redox biology emphasizes the advantageous signaling functions of ROS at low concentrations, whereas oxidative stress is a consequence of excessive ROS leading to molecular injury (Mittal et al., 2022).

Denham Harman's hypothesis regarding aging postulates that ROS play a role in the aging process by harming essential cellular macromolecules, particularly within mitochondria. Malfunctioning mitochondria generate an excess of ROS, resulting in a gradual decline of cellular functionality over time, Fig. 3, B (Giorgio et al., 2007).

The process of aging speeds up when the repair of tissues struggles due to insufficient elimination and replacement of damaged cells by stem cells. The upkeep of stem cells hinges on their capacity to self-replicate and differentiate, exemplified by hematopoietic stem cells (HSCs), which perpetually renew blood cell lineages. Nevertheless, excessive levels of ROS can hinder stem cell functionality by blocking self-renewal, even without directly inflicting cellular injury. Therefore, maintaining

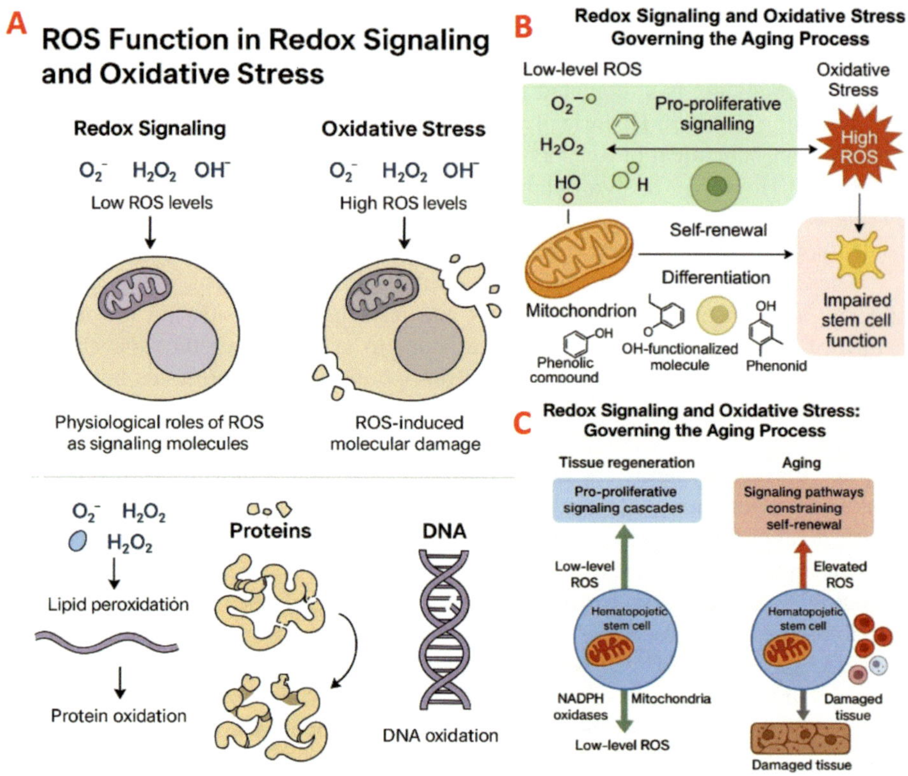

Fig. 3. The schematic of ROS reactions in (A) oxidative stress, (B) aging process in mitochondria, (C) regeneration and aging in HSCs.

a balance in ROS signaling is vital for robust stem cell activity, Fig. 3, C (Ludin et al., 2014).

3. Redox Polymers in Energy Storage: Emerging Paradigms for Electrochemical Accumulation

3.1 Fundamental Principles of Electrochemical Cells

A rudimentary electrochemical cell, commonly termed a battery, is constituted by a bipartite electrode assembly segregated by a nanoporous diaphragm and an ionically conductive electrolyte, facilitating inter-electrode ionic flux. An external electrical circuit permits the vectorial migration of electrons between these electrochemical interfaces. During the galvanostatic charging phase, the anodic pole undergoes an oxidative transformation, while the opposing cathodic pole experiences reductive processing. This electrochemical transduction enables the cell to store electrical charge via Faradaic mechanisms, necessitating an energy input to impel electrons towards the electropositive terminal, thereby enacting the charging process. Conversely, during the discharge cycle, the stored electrochemical potential is reconverted into electrical energy as electrons traverse towards the electronegative terminal. The physicochemical transformations transpiring at the respective electrodes are inherently redox processes, with each electrode material exhibiting a characteristic electrochemical potential at which its constituent reactions are thermodynamically favored. Ergo, the electromotive force (voltage) attainable from the cell is a direct function of the differential in the redox potentials of the constituent electrode materials. An additional salient characteristic of the electrochemical cell is its specific charge storage capacity, which quantifies the magnitude of electrical charge that can be sequestered per unit mass of the electroactive material. Consequently, the specific capacity is intrinsically linked to the stoichiometry of the electrode reaction, specifically the number of electrons transferred per molecular entity of the electrode material relative to its molar mass (Gu et al., 2024).

3.2 Organic Electrode Architectures for Electrochemical Energy Storage

Several molecular classes have demonstrated promise as constituents in organic electrochemical cells, encompassing organosulfur frameworks, stable free radicals, π-conjugated polymers (CPs), and redox-active small molecules. The electronic transport characteristics of CPs are intrinsically linked to their polyconjugated backbones. While organic materials are conventionally regarded as electrical insulators, the presence of extended π-electron delocalization in CPs engenders a discernible electronic band structure, imparting semiconducting attributes analogous to their inorganic congeners. These macromolecular architectures can be rendered electroconductive through controlled redox processes (doping), leading to the generation of mobile charge carriers (polarons, bipolarons) and a subsequent transition to a conductive regime beyond a critical doping concentration. Nevertheless, the intrinsic doping mechanism inherent to CPs typically constrains the attainable charge density per repeating unit to a fractional range (circa 0.3–0.5 elementary charges). This fundamental limitation consequently imposes a ceiling on the achievable gravimetric capacity of most CPs. Moreover, the absence of spatially confined electronic states in these systems results in redox potentials that exhibit a pronounced dependence on the dopant concentration, leading to a continuous modulation of cell voltage during electrochemical cycling, in contradistinction to the stable voltage plateaus observed with discrete redox-active molecular entities (Hammerich and Speiser, 2016).

In contradistinction, electrochemically active molecular entities, frequently leveraging carbonyl motifs as their redox locus, present precisely delineated and well characterized electrochemical reactivity profiles that are amenable to design via synthetic organic methodologies. Strategic installation of electron-modulating substituents in the proximate molecular scaffold enables fine-grained control over the electrochemical potential window. This inherent malleability endows them with considerable functional plasticity, permitting their implementation as either positive or negative electrode materials contingent upon the specific electrochemical device architecture. Notwithstanding this adaptability, such molecular species typically suffer from limited to negligible intrinsic electronic transport properties, necessitating the incorporation of conductive adjuncts within the electrode composite. Moreover, their inherent low molar mass engenders issues related to solute egress into the electrolytic medium, precipitating a progressive attenuation of charge storage capacity during sustained electrochemical cycling (De Riccardis, 2025; Fan et al., 2024; Hammerich and Speiser, 2016).

3.3 Conjugated Redox-Active Macromolecules: Multifunctional Architectures for Electrochemical Technologies

The strategic amalgamation of a charge-transporting polymeric main chain with pendant redox-functional units represents a compelling paradigm for circumventing numerous limitations inherent in conventional electroactive materials. The capacity for precise covalent tethering of tailored redoxophores to the polymer scaffold significantly broadens their applicability spectrum, extending well beyond conventional battery components. These Conjugated Redox Polymers (CRPs) can be rationally engineered for diverse technological domains, including electrochemical energy storage, electrochromic displays, photoresponsive devices, and advanced chemical sensors. Recent advancements have witnessed the integration of tailored redox polymers into both supercapacitor and redox flow battery systems. Notably, investigations have revealed that the incorporation of redox-active macromolecules can modulate the viscoelastic properties of electrolyte solutions, thereby impacting the fluid dynamics and pumpability of anolyte and catholyte streams within flow battery architectures (Gharahcheshmeh and Chowdhury, 2024).

3.4 Diverse Architectures of Organic Redoxophores: Molecular Transducers in Electrochemical Systems

Organic redox moieties, the functional heart of electroactive organic materials (comprising a carbonaceous framework), are molecular entities capable of engaging in reversible electron transfer processes, oscillating between oxidized and reduced states. This intrinsic electroactivity underpins their utility across a spectrum of technological applications, notably within energy storage and conversion platforms (Lee et al., 2020).

The vast expanse of organic chemistry provides access to a diverse library of functional groups and molecular architectures exhibiting reversible redox characteristics. Salient examples include:

Quinonoid Structures: These cyclic dicarbonyl compounds, characterized by conjugated π-electron systems, undergo a two-electron reductive transformation to their hydroquinone counterparts. Archetypal examples encompass benzoquinone, anthraquinone, and their structurally diverse derivatives, which are extensively investigated as electroactive components in batteries and supercapacitors.

Persistent Organic Radicals (Exemplified by TEMPO): Sterically hindered and electronically stabilized organic radicals, such as 2,2,6,6-tetramethylpiperidine-1-oxyl (TEMPO), can participate in single-electron redox transformations.

Macromolecular architectures incorporating TEMPO units find application in organic radical batteries.

Conjugated Polymers: Polymeric systems featuring alternating single and double bonds (e.g., polyaniline, polypyrrole, polythiophene) can be rendered electronically conductive through electrochemical or chemical doping (oxidation or reduction). In these materials, the polymer backbone itself serves as the electroactive locus.

Aromatic Systems with Electron-Modulating Substituents: Molecular entities such as viologens (bispyridinium salts) exhibit reversible single- or dual-electron reduction processes. The strategic introduction of electron-withdrawing or electron-donating functional groups on the aromatic core allows for precise modulation of their electrochemical potentials.

Organosulfur Motifs: Chemical functionalities incorporating sulfur atoms, such as thiolates, disulfides, and thiophenes, can participate in redox transformations involving changes in the sulfur oxidation state.

Nitroxide Radical Analogues: Similar to the prototypical TEMPO radical, a broader class of nitroxide radicals exhibits robust and reversible redox behavior.

Triarylamine Derivatives: These tertiary aromatic amines can undergo reversible one-electron oxidation, forming stable radical cation species.

Phosphine Oxides and Thioethers: These organochalcogenides demonstrate reversible electrochemical transformations at relatively negative potentials, making them attractive candidates for specialized electrochemical energy storage systems, particularly those designed for low-voltage operation.

Tetrazines: These nitrogen-rich heteroaromatic compounds undergo reduction half-reactions and are currently under investigation as promising electroactive components for next-generation energy storage technologies.

Catechols: These *ortho*-dihydroxybenzene derivatives can be electrochemically oxidized to their corresponding quinonoid structures and are frequently encountered as key structural motifs in bio-inspired polymeric materials developed for energy storage, contributing both redox functionality and, in certain cases, enhanced interfacial adhesion.

The selection of an appropriate organic redox moiety for a given electrochemical device is governed by a complex interplay of factors, including the desired electrochemical potential window, the kinetic and thermodynamic stability of the relevant redox couples, solute-solvent interactions influencing solubility, the molecule's theoretical charge storage capacity, and the performance specifications of the target energy storage or conversion system. The capacity for rational molecular design and synthesis to precisely tailor these electrochemical properties represents a significant advantage of employing organic materials in this field. In next sections we will introduce and discuss catechol-based redox polymers as organic redox moiety and their applications in energy storage systems.

4. Catechol-Based Redox Polymers

Redox Polymers can undergo reversible oxidation and reduction processes, which is why they are called electron exchange polymers (Yuan and Minteer, 2019). The electroactive sites located in the chemical structure of redox polymers are responsible for this process (Casado and Mecerreyes, 2020). Nitroxides (Boujioui and Gohy, 2019), quinones (Symons, 2021), phenothiazines (Golriz et al., 2015), and viologens (Chandra et al., 2023) are examples of electroactive moieties incorporated into the polymer's chemical structure to introduce redox capability. Quinone-based redox polymers have attracted much attention in recent years, due to their outstanding physicochemical and electrochemical versatility. Furthermore, quinone moieties can be readily obtained from bio-inspired polymers such as poly(dopamine), poly(catechol), lignin, and other sources (Gallastegui et al., 2022; Patil and Marcilla, 2020a). Due to their remarkable diversity, catechol-based systems have found an ever-growing number of applications (Sedó et al., 2013), such as bio-inspired adhesives (Saiz-Poseu et al., 2019), biomedical applications (Su et al., 2011), coatings (antimicrobial, antifouling, and anticorrosion) (Razaviamri et al., 2021), environmental (Joshi et al., 2020), and energy-related applications (Patil and Marcilla, 2020a), etc.

Polymers that incorporate the catechol group are utilized in numerous applications, including biocompatible glues, functional coatings and scaffolds, energy storage solutions, and lubricants (Sedó et al., 2013a; Stockdale et al., 2025; Zhang et al., 2018). Catechol can promote both reversible and irreversible attachment to surfaces, stemming from its ability to engage in various molecular interactions, such as complexation with metal ions, coordination to metal oxide surfaces, mono- and bi-dentate hydrogen bonding, π-π stacking, cation-π interactions, and the creation of reversible boron–oxygen covalent bonds (Li et al., 2010; Shabeer A. Mian et al., 2010; Shabeer Ahmad Mian et al., 2014; Patil et al., 2018). These attributes are greatly sought after for the creation of innovative and sophisticated multifunctional biomimetic substances with exceptional characteristics.

4.1 Interactions and Reactions of Catechol

Nature is a big source of catechol and its derivatives. For example Dopamine, as one of the famous catechol derivatives is responsible for the versatile adhesion of mussels, Fig. 4 (Pinnaratip and Lee, 2021; Razaviamri et al., 2021).

As shown in Fig. 5, catechol can undergo three main reactions: as an antioxidant, as a chelating agent and as a radical scavenger (Faure et al., 2013).

The most important reactions of catechols and ortho-quinones are shown in Fig. 6. In summary, catechol moieties exhibit the ability to undergo both reversible noncovalent (hydrogen bonding, π–π electron interaction, cation–π interaction, coordination with metal oxide surfaces and metal ions) and covalent (catechol–boronate complexation, polymerization, and irreversible bonding to organic

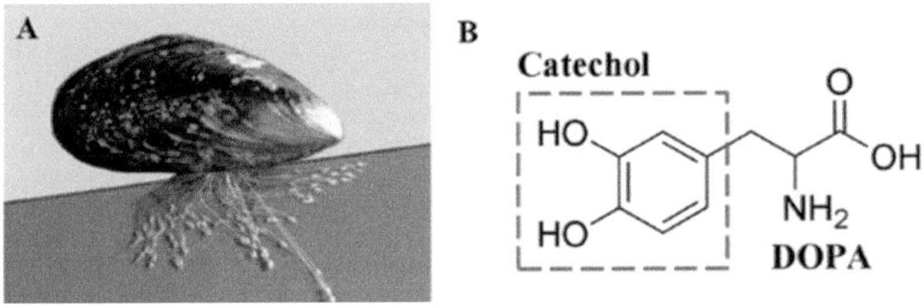

Fig. 4. Photograph of a marine mussel, Mytilus californianu (A). Chemical structure of DOPA with a catechol side chain (B) (Pinnaratip and Lee, 2021) Published by American Chemical Society.

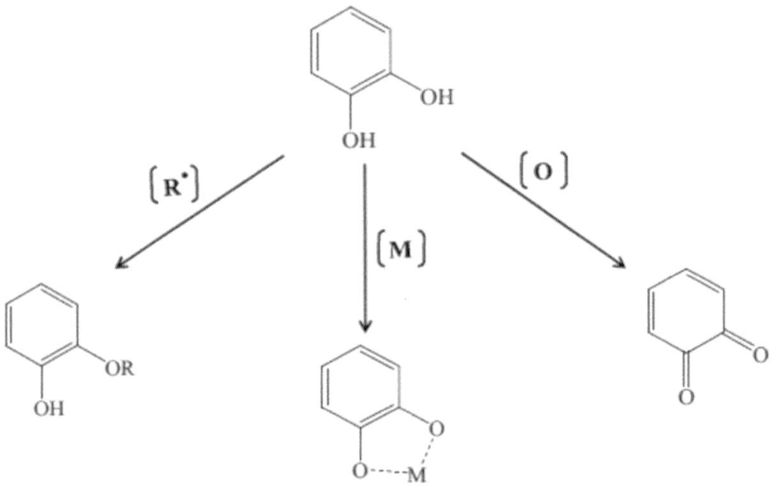

Where R• means Radical; M, Metal and O, Oxidant

Fig. 5. Main chemical properties of catechols (Faure et al., 2013).

substrates or molecules bearing, -thiol, -amine, and -imidazole functional groups) interactions. The oxidation state and, therefore, the catechol's physicochemical properties strongly depend on the surrounding pH. In an acidic medium, the catechol exhibits strong adhesion, while increasing pH leads to catechol oxidation and adhesion loss (Pinnaratip and Lee, 2021; Razaviamri et al., 2021). Introducing electron-donating and electron-withdrawing functional groups affects the oxidation tendency (Ding et al., 2015).

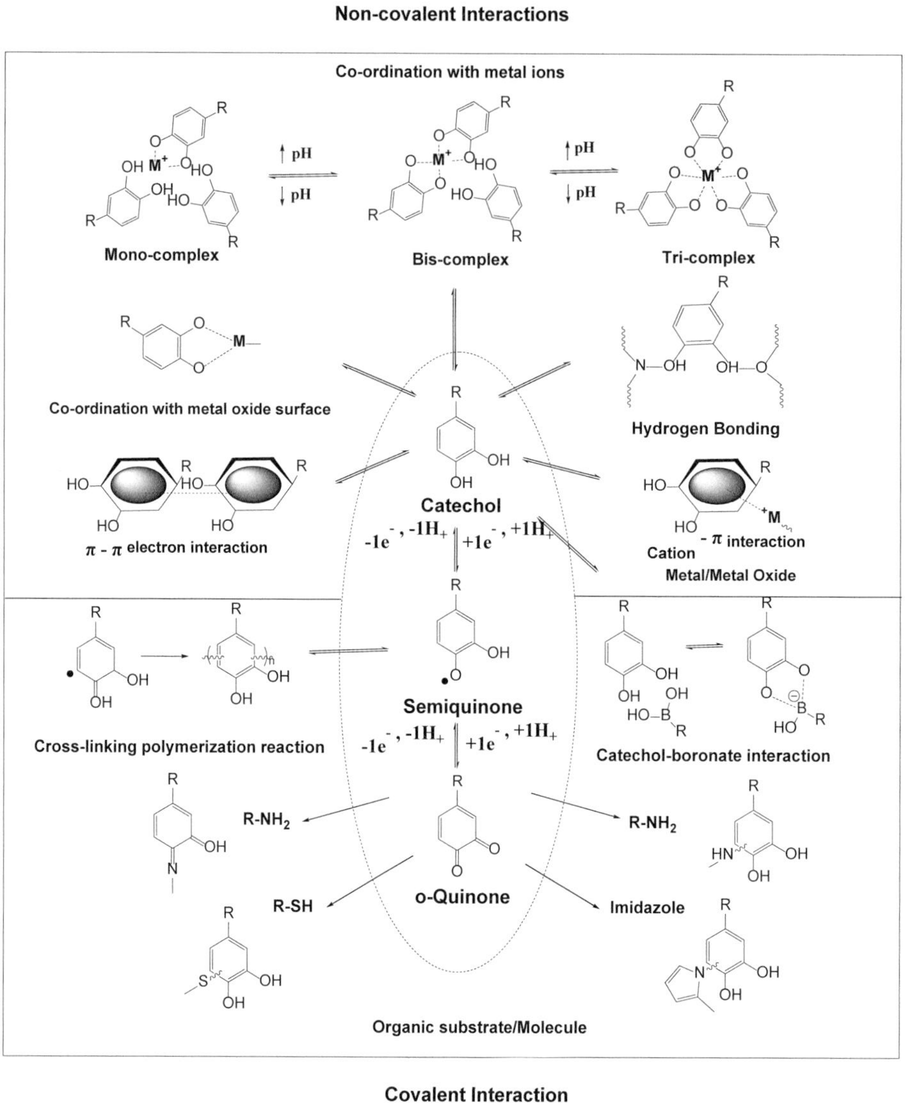

Fig. 6. Possible interactions and reactions of catechol, semiquinone, and ortho-quinones.

4.2 Design and Synthesis of Catechol-Functionalized Polymers

In recent years, significant advancements have been achieved in the meticulous design of innovative polymers featuring catechol groups, especially those synthesized through the polymerization of monomers with pendant catechol functionalities. The meticulous application and integration of catechol protection/deprotection reactions alongside regulated polymerization methods has truly paved the way for the creation of groundbreaking well-structured (co)polymers boasting adjustable composition, functionality, and architecture. This large-scale molecular engineering has enhanced the comprehension of the crucial structural elements that govern the characteristics of polymers, and proved to be an effective instrument for the ideal formulation of the material that seamlessly aligns with the intended application (Patil et al., 2018). Versatile, reactive, reversible, pH-sensitive polymers can be designed and synthesized by incorporating the catechol moiety into the polymer backbone, side, or end chain that is demonstrated in Fig. 7 (Faure et al., 2013).

One common example of introducing the catechol moiety into the polymer backbone is the oxidative self-polymerization of dopamine in a weakly alkaline environment, which results in the formation of a polydopamine film on the surface of different substrates, which introduces inherent adhesion and redox activity of catechol (Fig. 6) (Du et al., 2020; Li et al., 2022). The oxidative polymerization of catechol and its analogs has been extensively utilized due to the chemical reactivity of catechol precursors, the simplicity of polymerization, and the strong coating tendency driven by polymerization-induced substrate independence (Lee et al., 2007; Liu et al., 2014). Managing the partial oxidation process is essential for enhancing the electrochemical characteristics of the polydopamine derivative. As shown in Fig. 8, catechol-based coating could be further used for desired surface functionalization (Faure et al., 2013).

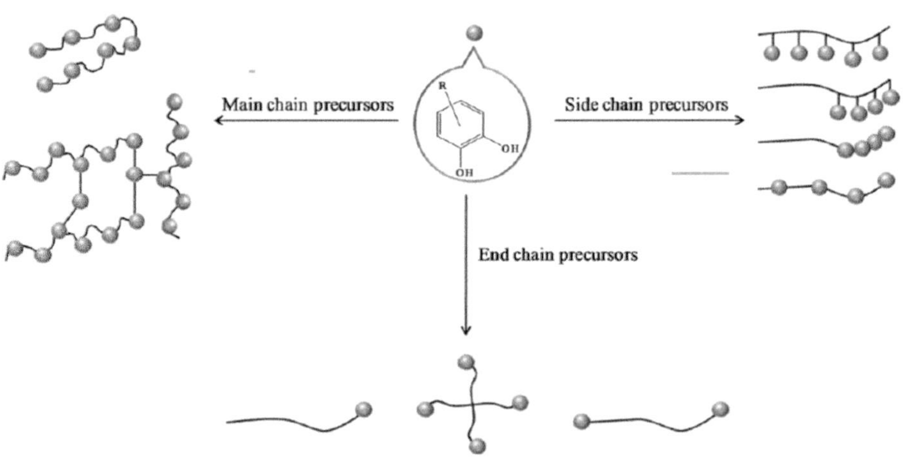

Fig. 7. Incorporation of catechol building blocks into the polymer (Faure et al., 2013). 6023821004471.

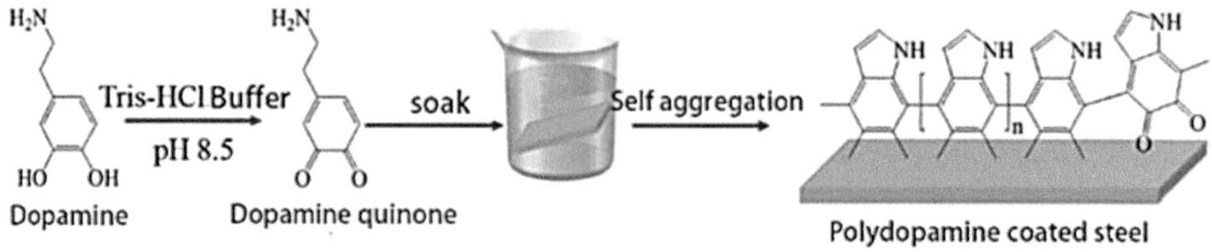

Fig. 8. Preparation process of polydopamine coating on steel surface (Li et al., 2022).

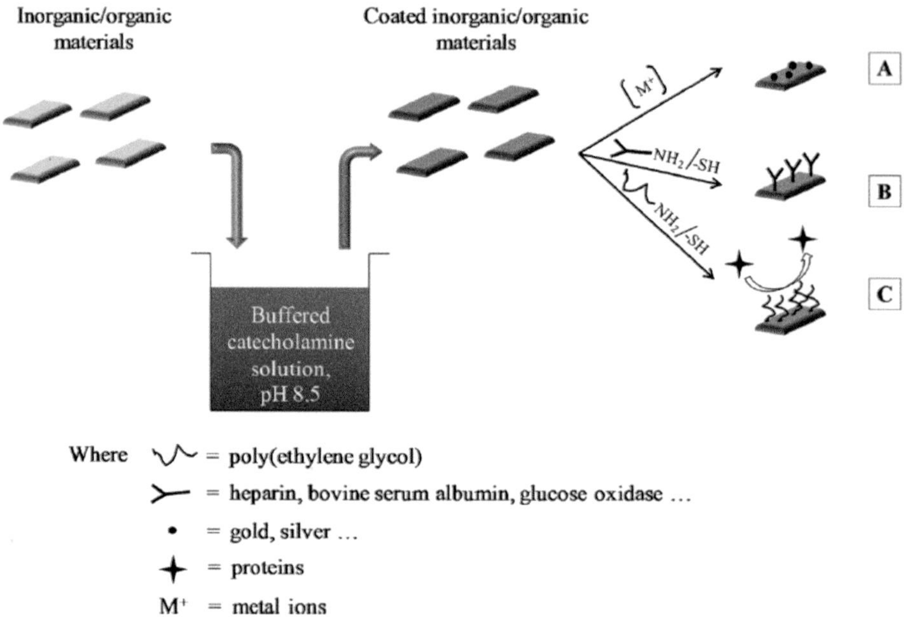

Fig. 9. Polydopamine coating and further derivatizations (Faure et al., 2013). 6023821004471.

Catechol units could be grafted onto chain ends by chemical ligation, usually to obtain materials for metal nanoparticle stabilization in the biomedical field. Metallic ions can be integrated to enhance the stability of catechol-derived polymers and boost their electrochemical effectiveness (Fig. 9, A). Among various methods, the chemical attachment of catechols to pre-existing (bio)polymer frameworks has emerged as the most acknowledged technique. Certainly, biopolymers are found in an extensive assortment of shapes, showcasing many active functional groups including $-NH_2$, –COOH, –CHO, and –OH, among others (Fig. 9, B). Moreover, these functional groups can be readily integrated into (semi)synthetic polymers (Fig. 9, C) using widely recognized synthetic methods and can be easily modified with a variety of catechol derivatives containing complementary functional groups through the creation of amide, ester, and various linkages either at the terminal ends or as side chain precursors.

Monomers bearing catechol functional side groups, such as DOPA methacrylamide, can form catechol side chain polymers. Puertas-Bartolomé et al.

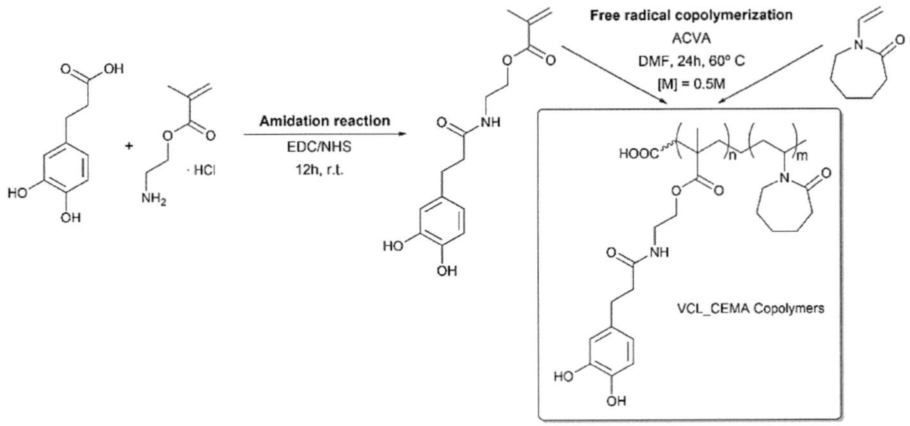

Fig. 10. Synthesis and characterization of the long-arm catechol-functionalized monomer and its homopolymer (Puertas-Bartolomé et al., 2018). 6023841207794.

(Puertas-Bartolomé et al., 2018) synthesized a bioadhesive based on copolymers of a flexible catechol-functionalized methacrylate (CEMA) and N-vinylcaprolactam (Fig. 10).

Messersmith et al. (Lee et al., 2002) synthesized linear and branched DOPA modified poly(ethylene glycol)s (PEG-DOPAs) containing one to four DOPA end groups. The PEG macromers altered with DOPA modified oxidative cross-linking, aided by sodium periodate, resulting in the creation of hydrogels. The capacity to regulate gelation rates and physical characteristics enables tailoring to particular application requirements.

The Reversible Addition-Fragmentation Chain Transfer (RAFT) technique was mainly used to polymerize catechol-derived vinylmonomers in their protected form. The RAFT polymerization method was similarly utilized to produce cutting-edge single-ion conducting redox-active copolymers, P(DA-stat-LiAMPS) and P(4VC-stat-LiSS), via the polymerization of their constituent comonomers, followed by deprotection and H+/Li+ exchange processes, which were subsequently employed as cathode materials for lithium-ion storage (Patil et al., 2017). The regulated characteristics of the RAFT method were utilized to create precisely structured block copolymers featuring catechol groups.

Detrembleur et al. (Patil et al., 2016) studied the first formulation of innovative catechol-inspired poly(ionic liquid)s (PILs), showcasing various counteranions through Cobalt-facilitated radical polymerization (CMRP) of safeguarded monomers, **123** or **124**, utilizing an alkylcobalt (III) complex as the initiator and mediating agent, succeeded by a sequence of acid-catalyzed deprotection and counteranion exchange procedures. PILs derived from multifunctional polyphenols provide a flexible platform for creating advanced surfaces and functional materials for electrochemical energy storage. This is due to the combined advantages of polyphenols—such as catechol and pyrogallol—which offer both redox activity and strong adhesion, along with the customizable properties of PILs.

Gallastegui et al. (Gallastegui et al., 2022) present a novel synthesis method for catechol-based redox-active polymer nanoparticles (cRPNs) and investigates their aging behavior and electrochemical performance. The cRPNs were produced through a single-step emulsion polymerization of dopamine methacrylamide (DMA) in water at a temperature of 70°C, employing 2,2'-azobis(2-methylpropionamidine) dihydrochloride (AIBA) as the initiator. This method facilitated the creation of nanoparticles with adjustable dimensions varying between 25 and 165 nm, all without requiring the safeguarding of the catechol groups. This is important due to the possible disruption that catechols may cause in radical polymerization processes.

Table 1. Advantages and disadvantages of polymerization.

Synthesis method	Advantages	Disadvantages
Plasma-Assisted Deposition	✓ Low-temperature processing (Fridman, 2008) ✓ Solvent-free process (Chan et al., 1996) ✓ Environment friendly (Chan et al., 1996) ✓ Customizable chemical functionality (Friedrich, 2012) ✓ Conformal coating & strong adhesion (Friedrich, 2012; Yasuda and Matsuzawa, 2005)	✓ Low throughput & complexity (Wertheimer et al., 2005) ✓ Elevated equipment and functionality cost (Nema and Jhala, 2015) ✓ Structural limitations (Biederman et al., 2004; Zeuner and Biederman, 2004) ✓ Film thickness (Biederman et al., 2004; Zeuner and Biederman, 2004) ✓ Harm to substrate surfaces (Baniya et al., 2021)
Enzymatic Polymerization	✓ Green & mild synthesis (Liu et al., 2014) ✓ Surface coating of selective oxidation (Ryu et al., 2018) ✓ Tunability & biocompatibility (Lynge et al., 2011)	✓ Limited polymerization & slow kinetics (Meskher and Achi, 2024) ✓ Scalability & stability challenges (Liu et al., 2014)
Metal Chelation Polymerization	✓ Green and simple synthesis (Lee et al., 2007) ✓ Reversible and strong Cross-Linking (Ejima et al., 2017) ✓ Adhesion Enhancement & mechanical robustness (Yan et al., 2021) ✓ Facilitated and high redox activity (Busqué et al., 2014)	✓ Metal Ion leaching (Dzhardimalieva and Uflyand, 2018) ✓ Lower conductivity without supplementation (Wang et al., 2022) ✓ pH sensitivity of chelation (Dzhardimalieva and Uflyand, 2018)
Controlled Radical Polymerization	✓ Precise control of molecular weight and architecture (Matyjaszewski and Xia, 2001) ✓ Functional monomer compatibility (Perrier, 2017) ✓ Scalability, reproducibility & solution processability (Zhang et al., 2018)	✓ Requirement for functional group protection/deprotection (Hassan et al., 2024) ✓ Limited electrochemical stability at high potentials (Perrier, 2017) ✓ Oxygen sensitivity (Perrier, 2017)
Protection/ Deprotection Polymerization	✓ Improves monomer stability (Jarowicki and Kocienski, 2001) ✓ Versatile deprotection options (Jarowicki and Kocienski, 2001)	✓ Risk of incomplete deprotection (Jarowicki and Kocienski, 2001) ✓ Reduced sustainability (Sartori et al., 2004)

Scientists use different methods to create redox polymers from catechol, and each method has its pros and cons. For example, enzymatic polymerization provides a more eco-friendly and milder approach by utilizing biological catalysts, steering clear of harmful chemicals and severe environments (Liu et al., 2014). While it's green and selective, it often faces issues like slow reaction speeds and a limited choice of monomers. Other advantage and disadvantages of synthesis methods are summarized in Table 1.
The schematic of synthesis methods are illustrated in Fig. 11.

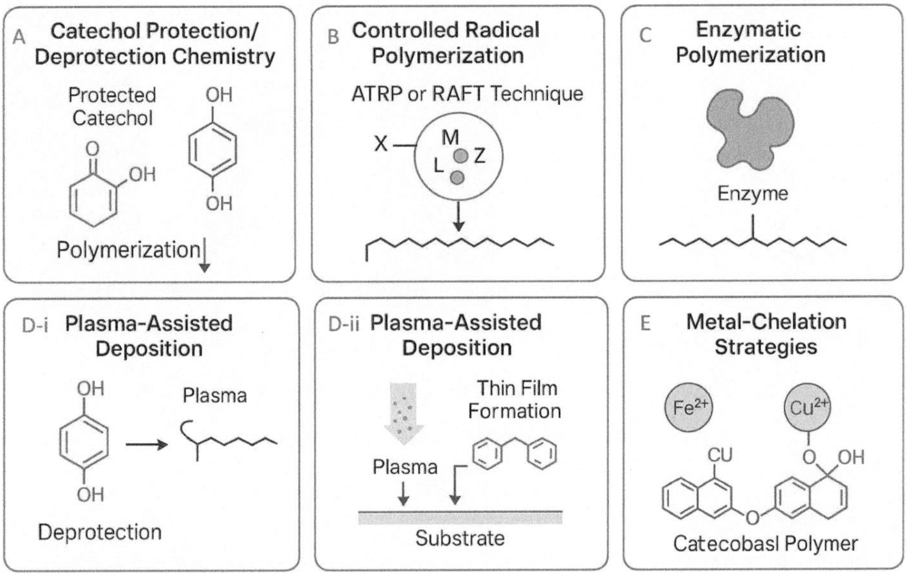

Fig. 11. The schematic of (A) protection/deprotection polymerization, (B) Controlled radical polymerization, (C) Enzymatic polymerization, (Di, ii) Plasma assisted deposition, (E) Metal chelation strategies.

5. Applications in Energy Storage Devices

Catechol-based redox polymers are experiencing huge growth in safe, sustainable, multifunctional, and bioinspired energy storage applications due to their significant, previously mentioned electrochemical properties, as well as their surface anchoring propensity and adhering capability (Patil et al., 2018). Catechol-functionalized polymers have been explored for use in aqueous rechargeable H^+, Li-ion, and post-Li-ion batteries (Patil et al., 2019) due to their high theoretical capacity (up to 496 mAh g^{-1}) and elevated redox potential of approximately 3.1 V (for ortho-quinones, compared to less than 2.8 V for para-quinones) versus Li/Li^+ (Wang et al., 2019; Zhu et al., 2019). Different electrochemical energy storage technologies are exhibited in Fig. 12. Capacitors, rechargeable static batteries, Lithium-ion batteries, and redox flow batteries are the most prominent applications related to catechol-based redox polymers in energy storage devices (Patil and Marcilla, 2020a).

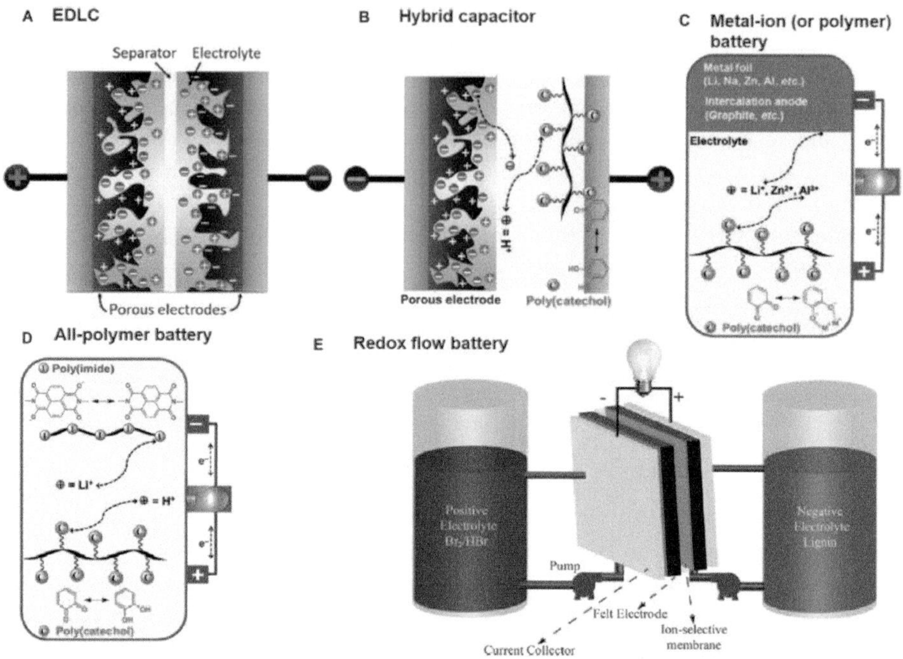

Fig. 12. Schematic illustration of different energy-storage devices described in this chapter. (A) Electric double-layer capacitor. (B) Hybrid capacitor. (C) Metal-ion (or polymer) battery. (D) All-polymer battery. (E) Redox flow battery (Patil and Marcilla, 2020a). 1607532.

5.1 Capacitors

In contrast to batteries, supercapacitors store energy through the physical adsorption of electrolyte ions onto electrodes with a high surface area, a mechanism known as electric double-layer capacitance (EDLC). As a result, supercapacitors exhibit several advantages that batteries currently cannot achieve, including higher power density, charging in less than a second, extremely long cycle life, and the ability to operate across a broad temperature range—though they store less energy compared to battery materials (Wang et al., 2012). Catechols capture considerable attention as cathode substances for energy storage, thanks to their remarkable capacity and redox potential. Due to electron acceptance, storage, and donation, catechol-based redox polymers offer redox-capacitor capabilities (Kim et al., 2017). Activated carbon electrodes are prepared via catechol polymers carbonization for application in EDLC. Pognon et al. (Pognon et al., 2012) exhibited that the redox activity of the surface-immobilized catechol groups, combines well with the electrochemical double layer capacitance of the high surface area black pearls carbon.

Poly(catechol)s, especially lignins, combined with usually two types of conductivity-enhancing additives and conductive polymers, are used to make electrodes for hybrid and pseudocapacitors. Dong et al. (Dong et al., 2023) introduce an innovative method for creating high performance supercapacitor electrodes by constructing lignin/polypyrrole (PPy) interwoven networks within cellulose nanofibril (LCNF) composite membranes that incorporate lignin. The LCNFs/PPy composites demonstrate a multi-layered porous architecture with an extensive surface area, enhancing effective ion movement and charge storage. Electrodes containing 11.4% lignin exhibited remarkable tensile strength (20.46 MPa) and outstanding areal specific capacitance (2567 mF cm^{-2}, 1 mA cm^{-2}). Therefore, electrodes exhibit outstanding rate performance and durability in cycling, owing to the collaborative impact of the conductive PPy network and the resilient carbon nanofiber structure.

Eumelanin-based electrodes are also assembled into aqueous pseudocapacitors. Due to the broadband absorption of eumelanin in the UV–Visible region, solar light could be utilized as an external trigger to improve the storage efficiency (Xu et al., 2019). Tannins, often extracted from tree bark, have also shown significant potential in the development of carbon-based electrodes for electrochemical capacitors (Hoseinizadeh et al., 2023).

In addition to poly(catechol)s derived from bioresources that demonstrate electrochemical performance, synthetic electrodes based on poly(catechol) are also attracting interest for use in supercapacitors. Pirnat et al. (Pirnat et al., 2019b) synthesized of redox-active polymer nanoparticles based on poly(vinyl catechol). Spherical polymer nanoparticles with different particle sizes were obtained by varying the surfactant concentration during polymerization.

Touze et al. (Touzé et al., 2018) utilized diazonium chemistry to attach O-protected catechol groups onto activated carbon electrodes. By adjusting the dimensions of the safeguarding groups on the catechol structures, they were able to influence the redox potential of the attached layers. The research revealed that enlarging the O-protective groups on the catechol derivatives resulted in an upward adjustment in the redox potential of the attached layers. This ability to tune is essential for enhancing the energy storage properties of supercapacitor electrodes.

5.2 Electrodes for Static Batteries

Carbon derived from natural biomass, like lignin, could be used in metal-ion batteries. However, their electrochemical energy-storage mechanism is similar to other carbon-based anodes, and their redox feature is lost (Patil and Marcilla, 2020a). The process of superlithiation, supersodiation, or superpotassiation of catechol polymers demonstrates high storage capacity and overcomes the limited storage challenge in metal-ion batteries based on carbon and transition metal oxide electrodes (Han et al., 2012). Coordination/non-coordination of metal cations in redox reactions paves the way for making organic cathodes in monovalent-ion (sodium and lithium ions), multivalent-ion (magnesium and zinc ions), and all-polymer (poly (catechol)s) batteries (Patil and Marcilla, 2020a).

A study (Pirnat et al., 2019) introduces an innovative method for the preparation of RPNs using poly(4-vinyl catechol) and poly(3-vinyl catechol). These RPNs were generated via a dual-stage procedure consisting of mini-emulsion polymerization and subsequent removal of protective groups from catechol monomers (Fig. 13, A, B). RPNs featuring 1% divinylbenzene (DVB) cross-linker exhibited peak electroactivity and durability during cycling (Fig. 13, C). An increased quantity of cross-linkers enhanced structural integrity but lowered electroactivity, whereas

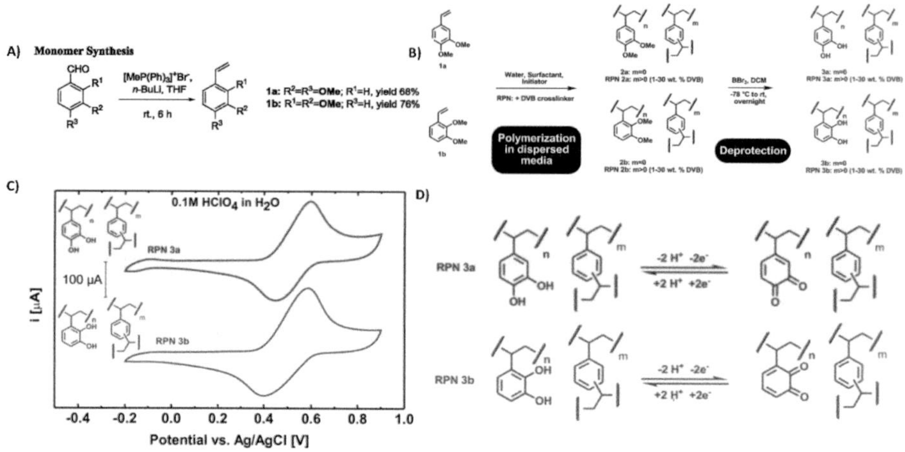

Fig. 13. (A) Wittig reaction for the synthesis of protected vinyl catechols, (B) reaction schematic for preparation of polymers 3a and 3b and redox active polymer nanoparticles RPN 3a and RPN 3b, cross-linked with DVB, (C) The CV analysis of RPN 3a and RPN 3b in a 0.1 M HClO4 electrolyte (1% DVB, 40 nm particle size, 3rd cycle, 20 mV/s, −0.2–0.9 V relative to Ag/AgCl), (D) diagrammatic illustration of the suggested redox process of RPNs in an acidic aqueous solution (Pirnat et al., 2019).

a decreased amount caused particle meld and impaired performance. The results emphasize the capability of these RPNs to serve as cathode components in organic batteries, presenting benefits like diminutive particle dimensions, reversible redox activity, and suitability for diverse electrolytes (Fig. 13, D).

Within the domain of rechargeable power sources, lithium-ion batteries (LIBs) have garnered substantial attention and undergone swift progress over the last four decades thanks to their superior performance capability and energy density, lightweight construction, durability, high discharge rates, and environmental sustainability (Nitta et al., 2015). These batteries include positive and negative poles of the cell caps, where the anode (the terminal of an electrochemical cell, negative pole) and cathode (the terminal of an electrochemical cell, positive pole) are separated by a permeable barrier (separator) that allows the movement of charge carriers (Li^+ ions in LIB) through the electrolyte medium. In the discharge phase, the anode undergoes oxidation while Li^+ ions are infused into the cathode, accompanied by the flow of electrons from the anode to the cathode via an external circuit (Muench et al., 2016).

Patil and colleagues (Patil et al., 2017) present an innovative category of catechol-modified polymers aimed at serving as superior cathode materials for LIBs. The ortho-quinone variant of catechol demonstrates a redox potential of approximately 3.1 V relative to Li/Li^+, surpassing the para-quinones which are below 2.8 V. This elevated potential enhances the energy yield in battery systems. This study offers an enticing blend of substantial capacity, enhanced redox potential, robust structural integrity, and eco-friendly sustainability, establishing them as strong contenders for the cathodes of future LIBs.

Sun and colleagues (Sun et al., 2016) developed the electrode substance via the oxidative polymerization of dopamine, which was subsequently partially oxidized to achieve a balance between the amount of redox-active carbonyl groups and the material's structural integrity and conductivity. The enhanced polydopamine derivative displayed outstanding electrochemical efficacy, providing an impressive capacity of 1818 mAh g^{-1} for LIBs and 500 mAh g^{-1} for sodium-ion batteries (SIBs). Remarkably, the material obtained from polydopamine served proficiently as active electrode component and the adhesive, removing the necessity for extra binding substances and streamlining the electrode manufacturing process.

5.3 Redox Flow Battery

Replacement of toxic, expensive, vanadium salt-based electrolytes by environmentally friendly and abundant organic-based redox flow batteries is gaining increasing attention due to their potential for sustainable, low-cost, safer, and metal-free energy storage applications (Adeniran et al., 2022; Patil and Marcilla, 2020a). Mukhopadhyay et al. (Mukhopadhyay et al., 2018) first proved the potential of lignin as an abundant biopolymer and extremely low-cost electrolyte in a redox flow battery. The scientists utilized ultrafiltration methods to refine lignosulfonate, a soluble variant of lignin extracted as a secondary product from paper and biofuel sectors. This approach successfully eliminated low-molecular-weight contaminants, producing a refined lignin solution abundant in redox-active quinone groups. The refined lignosulfonate was subsequently utilized as the anolyte in the flow battery device. In conjunction with a bromine-derived catholyte (Br_2/Br^-), the flow battery attained current densities reaching as high as 20 mA/cm^2 and exhibited a charge polarization resistance of 15 $\Omega \cdot cm^2$. Table 2 provides a summary of contemporary

Table 2. Advances in catechol-based redox polymers for energy storage.

Type of Catechol-based Redox Polymer	Synthesis Method	Gravimetric Capacities	Application	Ref.
Poly (catechol) homopolymer and copolymer	Controlled radical polymerization	180 mAh/g	Aqueous zinc polymer batteries	(Patil et al., 2021)
Catechol-rich plasma deposited polymer	Atmospheric plasma deposition	Enhanced stability and conductivity	Lithium ion batteries	(Aqil et al., 2021)
Catechol-functionalized polymer composites	Plasma-assisted deposition	High gravimetric capacity exceeding Li-ion batteries	Lithium-sulfur batteries	(Hu et al., 2017)
Catechol-pyridine redox polymer	Metal-chelation method	Improved energy storage capacity	Organic electrodes for lithium batteries	(Gallastegui et al., 2020)
Sulfonated polycatechol-PEDOT composite	Conductive polymer-based energy storage	95 mAh/g (thin film), 178 F/g (nanocomposite)	Morphology optimization via controlled deposition	(Vereshchagin et al., 2022)
Poly (catechol) cathode	Aqueous all polymer batteries	80.6 Wh/kg (energy density), 348 kW/kg (power density)	Polymerization with optimized electrolyte composition	(Patil et al., 2021)

studies on catechol-infused redox polymers, emphasizing their varied uses in energy storage. In general, the results demonstrate the promise of catechol-derived polymers to propel the advancement of eco-friendly and high-efficiency electrochemical technologies.

6. Advantages and Disadvantages of Catechol-based Redox Polymers in Energy Storage Devices

Redox polymers derived from catechol are becoming notable materials for energy storage thanks to their flexibility, sustainability, and electrochemical efficiency (Patil, Aqil et al., 2018). Sourced from biological materials, they present a more eco-conscious choice compared to traditional substances, which frequently depend on limited or harmful components like lithium and cobalt (Silverman and Roberto, 2007). Their capacity to partake in reversible oxidation-reduction reactions ensures exceptional efficiency, while their adaptability enables tailored characteristics for various applications. Moreover, these polymers are safe and non-toxic, mitigating the dangers linked to thermal runaway or harmful byproducts in battery technologies (Patil, Jerome, et al., 2018). Additionally, their possible economic efficiency renders them a compelling choice for upcoming energy strategies.

Nonetheless, redox polymers derived from catechol still possess disadvantages. A major obstacle is enduring stability, as they might be prone to aging influences that could change their redox potential and general efficiency. Scalability continues to be a challenge, since extensive production and market feasibility necessitate additional enhancement (Patil and Marcilla, 2020b). Certain catechol-derived substances could experience electrode dissolution, affecting their longevity throughout prolonged cycles (Lukyanov et al., 2021; Patil and Marcilla, 2020b). Redox-active polymer nanoparticles offer a promising approach to prevent the unwanted dissolution of organic electrode materials, while maintaining well-defined redox behavior (Gallastegui et al., 2022). Furthermore, despite demonstrating encouraging electrochemical characteristics, further refinement is required to boost effectiveness and guarantee real-world use in energy storage technologies (Patil and Marcilla, 2020b). Scientists are persistently investigating methods to enhance these substances, striving to mitigate their drawbacks while harnessing their advantages for eco-friendly and effective energy storage alternatives.

7. Conclusion and Perspectives

Due to the reversible redox activity and versatility of the catechol moiety, catechol-based redox polymers demonstrate a significant and promising capability in energy storage applications. Molecular design leads to the development of novel materials for high-performance capacitors, static battery electrodes, and redox flow batteries. The molecular architectural capabilities, redox reversibility, and tunable electronic properties of catechol-functionalized polymers lead to improved device efficiency and open pathways for sustainable, flexible, versatile, and lightweight energy storage systems.

Looking ahead, the future landscape of this field is shaped by several challenges and opportunities. Key areas for further research include enhancing the long-term stability of catechol-based polymers, reducing solubility losses in electrolyte environments, and developing scalable, cost-effective synthesis methods. Moreover, the integration of catechol-based polymers with emerging technologies such as flexible electronics, wearable devices, and hybrid energy systems offers exciting possibilities. Advances in molecular engineering, polymer chemistry, and device fabrication will be critical in fully harnessing the potential of catechol-based redox polymers. These materials could significantly evolve next-generation, sustainable energy storage solutions with continued interdisciplinary efforts.

References

Adeniran, A., Bates, A., Schuppert, N., Menon, A. and Park, S. 2022. Recent advances in aqueous redox flow battery research. Journal of Energy Storage 56: 106000. doi:https://doi.org/10.1016/j.est.2022.106000.

Aqil, A., Ouhib, F., Detrembleur, C. and Moreno-Couranjou, M. 2021. Atmospheric plasma deposition of bioinspired catechol-rich polymers: a promising route for the simple construction of redox-active thin films. Materials Advances 2(4): 1248–1252.

Baniya, H.B., Guragain, R.P. and Subedi, D.P. 2021. Cold atmospheric pressure plasma technology for modifying polymers to enhance adhesion: A critical review. Progress in Adhesion and Adhesives 6: 841–879.

Biederman, H., Kudrna, P. and Slavinska, D. 2004. Hard plasma polymers, composites and plasma polymer films prepared by rf sputtering of conventional polymers. Plasma Polymer Films 289–324.

Boujioui, F. and Gohy, J.-F. 2019. Ion-conducting redox-active polymer gels based on stable nitroxide radicals. Polymers 11(8). Retrieved from doi:10.3390/polym11081322.

Busqué, F., Sedó, J., Ruiz-Molina, D. and Saiz-Poseu, J. 2014. Catechol-based biomimetic functional materials and their applications. Bio-and Bioinspired Nanomaterials 277–308.

Casado, N. and Mecerreyes, D. 2020. Introduction to redox polymers: classification, characterization methods and main applications. *In*: Casado, N. and Mecerreyes, D. (eds.). Redox Polymers for Energy and Nanomedicine (pp. 0): The Royal Society of Chemistry.

Chan, C.-M., Ko, T.-M. and Hiraoka, H. 1996. Polymer surface modification by plasmas and photons. Surface Science Reports 24(1-2): 1–54.

Chandra, S., Lielpetere, A. and Schuhmann, W. 2023. Designing a high-potential metal-free viologen-based redox polymer for effective wiring of FAD-dependent glucose dehydrogenase. Sensors and Actuators B: Chemical 397: 134660. doi:https://doi.org/10.1016/j.snb.2023.134660.

Darmayanti, M.G., Tuck, K.L. and Thang, S.H. 2024. Carbon dioxide capture by emerging innovative polymers: status and perspectives. Advanced Materials 36(30): 2403324.

De Riccardis, M.F. 2025. Potential of conjugated polymers in nanostructured coatings for energy applications. pp. 511–533. In Multifunctional Nanostructured Coatings: Elsevier.

Ding, X., Vegesna, G.K., Meng, H., Winter, A. and Lee, B.P. 2015. Nitro-group functionalization of dopamine and its contribution to the viscoelastic properties of catechol-containing nanocomposite hydrogels. Macromolecular Chemistry and Physics 216(10): 1109–1119. doi:https://doi.org/10.1002/macp.201500010.

Dong, J., Li, P., Zeng, J., Wang, B., Gao, W., Xu, J. et al. 2023. Lignin/polypyrrole interpenetrating networks decorated Lignin-containing cellulose nanofibril composite membrane for high-performance supercapacitors. Chemical Engineering Journal 470: 144180. doi:https://doi.org/10.1016/j.cej.2023.144180.

Du, R., He, L., Li, P. and Zhao, G. 2020. Polydopamine-modified Al_2O_3/polyurethane composites with largely improved thermal and mechanical properties. Materials 13: 1772. doi:10.3390/ma13071772.

Dzhardimalieva, G.I. and Uflyand, I.E. 2018. Chemistry of Polymeric Metal Chelates (Vol. 257): Springer.

Ejima, H., Richardson, J.J. and Caruso, F. 2017. Metal-phenolic networks as a versatile platform to engineer nanomaterials and biointerfaces. Nano Today 12: 136–148.
Fan, K., Guan, L., Gu, Y., Liu, S. and Wang, C. 2024. Conjugated coordination polymers as multifunctional platform for electrochemical energy storage. Coordination Chemistry Reviews 519: 216098.
Faure, E., Falentin-Daudré, C., Jérôme, C., Lyskawa, J., Fournier, D., Woisel, P. et al. 2013. Catechols as versatile platforms in polymer chemistry. Progress in Polymer Science 38(1): 236–270. doi:https://doi.org/10.1016/j.progpolymsci.2012.06.004.
Fridman, A. 2008. Plasma Chemistry: Cambridge University Press.
Friedrich, J.F. 2012. The Plasma Chemistry of Polymer Surfaces: Advanced Techniques for Surface Design: John Wiley & Sons.
Gallastegui, A., Camara, O., Minudri, D., Goujon, N., Patil, N., Ruipérez, F. et al. 2022. Aging effect of catechol redox polymer nanoparticles for hybrid supercapacitors. Batteries & Supercaps 5(9): e202200155.
Gallastegui, A., Minudri, D., Casado, N., Goujon, N., Ruipérez, F., Patil, N. et al. 2020. Proton trap effect on catechol–pyridine redox polymer nanoparticles as organic electrodes for lithium batteries. Sustainable Energy & Fuels 4(8): 3934–3942.
Gharahcheshmeh, M.H. and Chowdhury, K.U.A. 2024. Fabrication methods, pseudocapacitance characteristics, and integration of conjugated conducting polymers in electrochemical energy storage devices. Energy Advances.
Giorgio, M., Trinei, M., Migliaccio, E. and Pelicci, P.G. 2007. Hydrogen peroxide: a metabolic by-product or a common mediator of ageing signals? Nature Reviews Molecular Cell Biology 8(9): 722–728.
Glass, P., Chung, H., Washburn, N.R. and Sitti, M. 2009. Enhanced reversible adhesion of dopamine methacrylamide-coated elastomer microfibrillar structures under wet conditions. Langmuir 25(12): 6607–6612.
Golriz, A.A., Suga, T., Nishide, H., Berger, R. and Gutmann, J.S. 2015. Phenothiazine-functionalized redox polymers for a new cathode-active material. RSC Advances 5(29): 22947–22950. doi:10.1039/C4RA17107A.
Gu, S., Chen, J., Hussain, I., Wang, Z., Chen, X., Ahmad, M. et al. 2024. Modulation of radical intermediates in rechargeable organic batteries. Advanced Materials 36(17): 2306491.
Hammerich, O. and Speiser, B. 2016. Organic Electrochemistry: CRC Press Boca Raton.
Han, X., Qing, G., Sun, J. and Sun, T. 2012. How many lithium ions can be inserted onto fused C6 aromatic ring systems? Angewandte Chemie International Edition 51(21): 5147–5151. doi:https://doi.org/10.1002/anie.201109187.
Hassan, M., Bhat, G.A. and Darensbourg, D.J. 2024. Post-polymerization functionalization of aliphatic polycarbonates using click chemistry. Polymer Chemistry 15(18): 1803–1820.
Hoseinizadeh, M., Salem, K.E., Gouda, A., Belanger, D. and Santato, C. 2023. Tannins for sustainable semi-solid-state supercapacitors. Waste and Biomass Valorization 14(10): 3475–3488. doi:10.1007/s12649-023-02125-4.
Hu, C., Kirk, C., Silvestre-Albero, J., Rodríguez-Reinoso, F. and Biggs, M.J. 2017. Free-standing compact cathodes for high volumetric and gravimetric capacity Li–S batteries. Journal of Materials Chemistry A 5(37): 19924–19933.
Jarowicki, K. and Kocienski, P. 2001. Protecting groups. Journal of the Chemical Society, Perkin Transactions 1(18): 2109–2135.
Joshi, S., Kathuria, H., Verma, S. and Valiyaveettil, S. 2020. Functional catechol–metal polymers via interfacial polymerization for applications in water purification. ACS Applied Materials & Interfaces 12(16): 19044–19053. doi:10.1021/acsami.0c03133.
Kim, E., Liu, Z., Liu, Y., Bentley, W. and Payne, G. 2017. Catechol-based hydrogel for chemical information processing. Biomimetics 2: 11. doi:10.3390/biomimetics2030011.
Lee, B.P., Dalsin, J.L. and Messersmith, P.B. 2002. Synthesis and gelation of DOPA-modified poly(ethylene glycol) hydrogels. Biomacromolecules 3(5): 1038–1047. doi:10.1021/bm025546n.
Lee, H., Dellatore, S.M., Miller, W.M. and Messersmith, P.B. 2007. Mussel-inspired surface chemistry for multifunctional coatings. Science 318(5849): 426–430.
Lee, S., Hong, J. and Kang, K. 2020. Redox-active organic compounds for future sustainable energy storage system. Advanced Energy Materials 10(30): 2001445.

Lee, Y.A., Lee, J., Kim, D.W., Yoo, C.-Y., Park, S.H., Yoo, J.J. et al. 2017. Mussel-inspired surface functionalization of porous carbon nanosheets using polydopamine and Fe^{3+}/tannic acid layers for high-performance electrochemical capacitors. Journal of Materials Chemistry A 5(48): 25368–25377.
Li, S.-C., Chu, L.-N., Gong, X.-Q. and Diebold, U. 2010. Hydrogen bonding controls the dynamics of catechol adsorbed on a TiO_2 (110) surface. Science 328(5980): 882–884.
Li, S., Guo, C., Wang, X., Guan, C. and Chen, G. 2022. Corrosion inhibition coating based on the self-assembled polydopamine films and its anti-corrosion properties. Polymers 14: 794. doi:10.3390/polym14040794.
Liu, Y., Ai, K. and Lu, L. 2014. Polydopamine and its derivative materials: synthesis and promising applications in energy, environmental, and biomedical fields. Chemical Reviews 114(9): 5057–5115.
Ludin, A., Gur-Cohen, S., Golan, K., Kaufmann, K.B., Itkin, T., Medaglia, C. et al. 2014. Reactive oxygen species regulate hematopoietic stem cell self-renewal, migration and development, as well as their bone marrow microenvironment. Antioxidants & Redox Signaling 21(11): 1605–1619.
Lukyanov, D., Apraksin, R., Yankin, A., Vlasov, P., Levin, O., Tolstopjatova, E. et al. 2019. Synthesis and electrochemical properties of poly (3, 4-dihydroxystyrene) and its composites with conducting polymers. Synthetic Metals 256: 116151.
Lukyanov, D.A., Vereshchagin, A.A., Soloviova, A.V., Grigorova, O.V., Vlasov, P.S. and Levin, O.V. 2021. Sulfonated polycatechol immobilized in a conductive polymer for enhanced energy storage. ACS Applied Energy Materials 4(5): 5070–5078.
Lynge, M.E., van der Westen, R., Postma, A. and Städler, B. 2011. Polydopamine—a nature-inspired polymer coating for biomedical science. Nanoscale 3(12): 4916–4928.
Ma, T., Easley, A.D., Thakur, R.M., Mohanty, K.T., Wang, C. and Lutkenhaus, J.L. 2023. Nonconjugated redox-active polymers: Electron transfer mechanisms, energy storage, and chemical versatility. Annual Review of Chemical and Biomolecular Engineering 14(1): 187–216.
Matyjaszewski, K. and Xia, J. 2001. Atom transfer radical polymerization. Chemical Reviews 101(9): 2921–2990.
Meskher, H. and Achi, F. 2024. Electrochemical sensing systems for the analysis of catechol and hydroquinone in the aquatic environments: a critical review. Critical Reviews in Analytical Chemistry 54(5): 1354–1367.
Mian, S.A., Saha, L.C., Jang, J., Wang, L., Gao, X. and Nagase, S. 2010. Density functional theory study of catechol adhesion on silica surfaces. The Journal of Physical Chemistry C 114(48): 20793–20800.
Mian, S.A., Yang, L.-M., Saha, L.C., Ahmed, E., Ajmal, M. and Ganz, E. 2014. A fundamental understanding of catechol and water adsorption on a hydrophilic silica surface: exploring the underwater adhesion mechanism of mussels on an atomic scale. Langmuir 30(23): 6906–6914.
Mittal, A., Vashistha, V.K. and Das, D.K. 2022. Recent advances in the antioxidant activity and mechanisms of chalcone derivatives: A computational review. Free Radical Research 56(5-6): 378–397.
Muench, S., Wild, A., Friebe, C., Haupler, B., Janoschka, T. and Schubert, U.S. 2016. Polymer-based organic batteries. Chemical Reviews 116(16): 9438–9484.
Mukhopadhyay, A., Hamel, J., Katahira, R. and Zhu, H. 2018. Metal-free aqueous flow battery with novel ultrafiltered lignin as electrolyte. ACS Sustainable Chemistry & Engineering 6(4): 5394–5400. doi:10.1021/acssuschemeng.8b00221.
Nema, S. and Jhala, P.B. 2015. Plasma Technologies for Textile and Apparel: CRC Press.
Nitta, N., Wu, F., Lee, J.T. and Yushin, G. 2015. Li-ion battery materials: present and future. Materials Today 18(5): 252–264.
Patil, N., Aqil, A., Ouhib, F., Admassie, S., Inganäs, O., Jérôme, C. et al. 2017. Bioinspired redox-active catechol-bearing polymers as ultrarobust organic cathodes for lithium storage. Advanced Materials 29(40): 1703373.
Patil, N., Aqil, M., Aqil, A., Ouhib, F., Marcilla, R., Minoia, A. et al. 2018. Integration of redox-active catechol pendants into poly (ionic liquid) for the design of high-performance lithium-ion battery cathodes. Chemistry of Materials 30(17): 5831–5835.
Patil, N., Cordella, D., Aqil, A., Debuigne, A., Admassie, S., Jérôme, C. et al. 2016. Surface-and redox-active multifunctional polyphenol-derived poly (ionic liquid)s: controlled synthesis and characterization. Macromolecules 49(20): 7676–7691.

Patil, N., Jerome, C. and Detrembleur, C. 2018. Recent advances in the synthesis of catechol-derived (bio) polymers for applications in energy storage and environment. Progress in Polymer Science 82: 34–91.

Patil, N., Mavrandonakis, A., Jérôme, C., Detrembleur, C., Palma, J. and Marcilla, R. 2019. Polymers bearing catechol pendants as universal hosts for aqueous rechargeable H+, Li-ion, and post-Li-ion (mono-, di-, and trivalent) batteries. ACS Applied Energy Materials 2(5): 3035–3041.

Patil, N. and Marcilla, R. 2020a. Catechol-containing polymers for electrochemical energy storage. *In*: Casado, N. and Mecerreyes, D. (eds.). Redox Polymers for Energy and Nanomedicine (pp. 0): The Royal Society of Chemistry.

Patil, N. and Marcilla, R. 2020b. Catechol-containing Polymers for Electrochemical Energy Storage.

Patil, N., Mavrandonakis, A., Jérôme, C., Detrembleur, C., Casado, N., Mecerreyes, D. et al. 2021. High-performance all-organic aqueous batteries based on a poly (imide) anode and poly (catechol) cathode. Journal of Materials Chemistry A 9(1): 505–514.

Patil, N., Palma, J. and Marcilla, R. 2021. Macromolecular engineering of poly (catechol) cathodes towards high-performance aqueous zinc-polymer batteries. Polymers 13(11): 1673.

Perrier, S. 2017. 50th Anniversary Perspective: RAFT Polymerization. A User Guide. Macromolecules 50(19): 7433–7447.

Pinnaratip, R. and Lee, B. 2021. Oxidation chemistry of catechol utilized in designing stimuli-responsive adhesives and antipathogenic biomaterials. ACS Omega, XXXX. doi:10.1021/acsomega.1c00006.

Pirnat, K., Casado, N., Porcarelli, L., Ballard, N. and Mecerreyes, D. 2019. Synthesis of redox polymer nanoparticles based on poly(vinyl catechols) and their electroactivity. Macromolecules 52(21): 8155–8166. doi:10.1021/acs.macromol.9b01405.

Pognon, G., Cougnon, C., Mayilukila, D. and Bélanger, D. 2012. Catechol-modified activated carbon prepared by the diazonium chemistry for application as active electrode material in electrochemical capacitor. ACS Applied Materials & Interfaces 4: 3788–3796. doi:10.1021/am301284n.

Puertas-Bartolomé, M., Fernández-Gutiérrez, M., García-Fernández, L., Vázquez-Lasa, B. and San Román, J. 2018. Biocompatible and bioadhesive low molecular weight polymers containing long-arm catechol-functionalized methacrylate. European Polymer Journal 98: 47–55. doi:https://doi.org/10.1016/j.eurpolymj.2017.11.011.

Razaviamri, S., Wang, K., Liu, B. and Lee, B. 2021. Catechol-based antimicrobial polymers. Molecules 26: 559. doi:10.3390/molecules26030559.

Razaviamri, S., Wang, K., Liu, B. and Lee, B.P. 2021. Catechol-based antimicrobial polymers. Molecules 26(3). doi:10.3390/molecules26030559.

Ryu, J.H., Messersmith, P.B. and Lee, H. 2018. Polydopamine surface chemistry: a decade of discovery. ACS Applied Materials & Interfaces 10(9): 7523–7540.

Saito, Y. and Yabu, H. 2015. Synthesis of poly (dihydroxystyrene-block-styrene)(PDHSt-b-PSt) by the RAFT process and preparation of organic-solvent-dispersive Ag NPs by automatic reduction of metal ions in the presence of PDHSt-b-PSt. Chemical Communications 51(18): 3743–3746.

Saiz-Poseu, J., Mancebo-Aracil, J., Nador, F., Busqué, F. and Ruiz-Molina, D. 2019. The chemistry behind catechol-based adhesion. Angewandte Chemie International Edition 58(3): 696–714. doi:https://doi.org/10.1002/anie.201801063.

Sartori, G., Ballini, R., Bigi, F., Bosica, G., Maggi, R. and Righi, P. 2004. Protection (and deprotection) of functional groups in organic synthesis by heterogeneous catalysis. Chemical Reviews 104(1): 199–250.

Sedó, J., Saiz-Poseu, J., Busqué, F. and Ruiz-Molina, D. 2013. Catechol-based biomimetic functional materials. Advanced Materials 25(5): 653–701. doi:https://doi.org/10.1002/adma.201202343.

Sedó, J., Saiz-Poseu, J., Busqué, F. and Ruiz-Molina, D. 2013a. Biomimetics: Catechol-based biomimetic functional materials (Adv. Mater. 5/2013). Advanced Materials 25(5): 792–792.

Sedó, J., Saiz-Poseu, J., Busqué, F. and Ruiz-Molina, D. 2013b. Catechol-based biomimetic functional materials. Advanced Materials 25(5): 653–701.

Silverman, H.G. and Roberto, F.F. 2007. Understanding marine mussel adhesion. Marine Biotechnology 9: 661–681.

Stockdale, H., Dowding, P.J., Wheeler, J.S., Vincent, B., Dalgliesh, R., Kenton, C. et al. 2025. Self-assembly of novel catechol-containing copolymers in a nonpolar solvent. Journal of Colloid and Interface Science 137737.

Su, J., Chen, F., Cryns, V.L. and Messersmith, P.B. 2011. Catechol polymers for pH-responsive, targeted drug delivery to cancer cells. Journal of the American Chemical Society 133(31): 11850–11853. doi:10.1021/ja203077x.

Sun, T., Li, Z.j., Wang, H.g., Bao, D., Meng, F.l. and Zhang, X.b. 2016. A biodegradable polydopamine-derived electrode material for high-capacity and long-life lithium-ion and sodium-ion batteries. Angewandte Chemie International Edition 55(36): 10662–10666.

Symons, P. 2021. Quinones for redox flow batteries. Current Opinion in Electrochemistry 29: 100759. doi:https://doi.org/10.1016/j.coelec.2021.100759.

Tchounwou, P.B., Yedjou, C.G., Patlolla, A.K. and Sutton, D.J. 2012. Heavy metal toxicity and the environment. Molecular, Clinical and Environmental Toxicology: Volume 3: Environmental Toxicology 133–164.

Touzé, E., Gohier, F., Daffos, B., Taberna, P.-L. and Cougnon, C. 2018. Improvement of electrochemical performances of catechol-based supercapacitor electrodes by tuning the redox potential via different-sized O-protected catechol diazonium salts. Electrochimica Acta 265: 121–130.

Vereshchagin, A.A., Potapenkov, V.V., Vlasov, P.S., Lukyanov, D.A. and Levin, O.V. 2022. Optimization of sulfonated polycatechol: PEDOT energy storage performance by the morphology control. Nanomaterials 12(11): 1917.

Waite, J.H. and Tanzer, M.L. 1981. Polyphenolic substance of Mytilus edulis: novel adhesive containing L-dopa and hydroxyproline. Science 212(4498): 1038–1040.

Wang, G., Zhang, L. and Zhang, J. 2012. A review of electrode materials for electrochemical supercapacitors. Chemical Society Reviews 41(2): 797–828.

Wang, H., Emanuelsson, R., Liu, H., Edstrom, K., Mamedov, F., Strømme, M. et al. 2019. Redox-state-dependent interplay between pendant group and conducting polymer backbone in quinone-based conducting redox polymers for lithium ion batteries. ACS Applied Energy Materials 2(10): 7162–7170.

Wang, W., Chen, F., Fang, L., Li, Z. and Xie, Z. 2022. Reversibly stretchable organohydrogel-based soft electronics with robust and redox-active interfaces enabled by polyphenol-incorporated double networks. ACS Applied Materials & Interfaces 14(10): 12583–12595.

Wertheimer, M., D'Agostino, R., Favia, P. and Oehr, C. 2005. Low temperature plasma processing of materials: past, present and future. Plasma Processes and Polymers 2(1): 7–15.

Xu, R., Gouda, A., Caso, M.F., Soavi, F. and Santato, C. 2019. Melanin: A greener route to enhance energy storage under solar light. ACS Omega 4(7): 12244–12251. doi:10.1021/acsomega.9b01039.

Yan, G., Chen, G., Peng, Z., Shen, Z., Tang, X., Sun, Y. et al. 2021. The cross-linking mechanism and applications of catechol–metal polymer materials. Advanced Materials Interfaces 8(19): 2100239.

Yasuda, H. and Matsuzawa, Y. 2005. Economical advantages of low-pressure plasma polymerization coating. Plasma Processes and Polymers 2(6): 507–512.

Yeo, J., Lee, J., Lee, S. and Kim, W.J. 2021. Polymeric antioxidant materials for treatment of inflammatory disorders. Advanced Therapeutics 4(4): 2000270.

Yu, M. and Deming, T.J. 1998. Synthetic polypeptide mimics of marine adhesives. Macromolecules 31(15): 4739–4745.

Yuan, M. and Minteer, S.D. 2019. Redox polymers in electrochemical systems: From methods of mediation to energy storage. Current Opinion in Electrochemistry 15: 1–6. doi:https://doi.org/10.1016/j.coelec.2019.03.003.

Zeuner, M. and Biederman, H. 2004. Corpuscular diagnostics of plasma polymerization processes. Plasma Polymer Films 85–142.

Zhang, H., Zhao, T., Newland, B., Liu, W., Wang, W. and Wang, W. 2018. Catechol functionalized hyperbranched polymers as biomedical materials. Progress in Polymer Science 78: 47–55.

Zhu, L., Ding, G., Xie, L., Cao, X., Liu, J., Lei, X. et al. 2019. Conjugated carbonyl compounds as high-performance cathode materials for rechargeable batteries. Chemistry of Materials 31(21): 8582–8612.

Index

P

R

S

T